Destination: Quality

How Our Small-Volume Building Firm Used TQM to Improve Our Business

by Gilbert J. Veconi,
with Charles A. Layne

Home Builder Press®
National Association of Home Builders
1201 15th Street, NW
Washington, DC 20005-2800
(800) 223-2665

This publication is designed to provide accurate and authoritative information in regard to the subject matter covered. It is sold with the understanding that the publisher is not engaged in rendering legal, accounting, or other professional service. If legal advice or other expert assistance is required, the services of a competent professional person should be sought.
—*From a Declaration of Principles jointly adopted by a Committee of the American Bar Association and a Committee of Publishers and Associations*

Destination: Quality
How Our Small-Volume Building Firm Used TQM to Improve Our Business

ISBN 0-86718-419-1

©1996 by Home Builder Press® of the
National Association of Home Builders of the United States

All rights reserved. No part of this book may be reproduced or utilized in any form or by any means, electronic or mechanical, including photocopying and recording or by any information storage and retrieval system without permission in writing from the publisher.

Cover by David Rhodes, Art Director, Home Builder Press

Printed in the United States of America

Library of Congress Cataloging-in-Publication Data
Veconi, Gilbert J., 1937-
  Destination—Quality : how our small-volume building firm used TQM to improve our business / Gilbert J.Veconi, Charles A. Layne.
    p.   cm.
  Includes bibliographical references.
  ISBN 0-86718-419-1
  1. Cannon Development (Rochester, N.Y.)—Management.
2. Construction industry—Management. 3. Total quality management—Case studies. I. Layne. Charles, A. 1943-   . II.Title.
HD9715.A2V43   1996   96-15881
624'.068'5—dc20   CIP

For further information please contact:
Home Builder Press®
National Association of Home Builders
1201 15th Street, NW
Washington, DC 20005-2800
(800) 223-2665

5/96 HBP/Kirby 1700
3/98 Reprint/Kirby 750

Contents

v Foreword
vii About the Authors
viii Acknowledgments

1 Introduction

8 Chapter One: Planning for the Journey
9 Was My Head On Straight?
14 Confirmation and Consensus
17 What Was Our Destination?
20 Final Preparations
27 We Were Off!

28 Chapter Two: First Days Out
29 Training Our Partners
40 Developing Written Material and Labor Specifications
45 What Were Buyers Telling Us?
50 How Long Were Our "First Days Out?"

52 Chapter Three: On the Road
53 Converting Customer Expectations into Process Improvement
76 The Value of Information
81 TQM Is an Attitude

83 Chapter Four: Sustaining the Journey
83 Keeping Our Engines Well Tuned
89 Staying on Course
93 Bumps in the Road
98 What's Left to Do?

100 Chapter Five: Are We There Yet?
100 Assessing Progress
111 Assessing Our Progress
117 What Have We Learned?
123 Just How Far Have We Come?

125 Resources

Foreword

Several years ago, when the NAHB Research Center was attempting to start its total quality management program for home builders, NAHB's Research Committee formed a quality subcommittee to provide guidance to the effort. The subcommittee's first meeting was held at the Washington Hilton during NAHB's spring board of director's meeting. We tried to explain the *quality process* to this group of successful builders who were sure that their homes were as good as could be possibly be built. Some talked of having strict standards that trade contractors had to follow. Others explained how they would keep going back, regardless of the cost, until the customer was totally satisfied. Others talked of quality as if it was a luxury for which customers are often unwilling to pay. Each was sure he or she understood quality and what was necessary to achieve it.

While the meeting was in progress, a man whom I had never seen before came into the room and sat in the back, obviously interested in the discussion. We continued to attempt to explain the difference between luxury and quality, the distinction between inspecting-in and building-in quality, and the need to have quality teams that include the trade contractors as well as the builder's own employees. Just as I was beginning to doubt my communication abilities, the man who came in late asked to speak. He introduced himself as Gil Veconi from suburban Rochester, New York.

Gil spoke of his quality process, how it was conceived, how it was perfected, and the benefits that were derived for his customers, employees, trade contractors, and his company, Cannon Development. As he spoke, the subcommittee members began to understand the notion that total quality is the result of management processes rather than inspection and supervision. I sat in admiration, not only of his commitment, but also of his ability to communicate the process in terms that builders could understand.

Since that day, the Research Center's Total Quality Construction program has taken off and now includes the National Housing Quality (NHQ) award, co-sponsored by *Professional Builder* magazine. One of the first-year award winners was—not surprisingly—Gil Veconi's Cannon Development. The award judges gave Cannon the highest point total of all the builders who applied. Because of his background and obvious expertise on the subject, Gil was invited to become an NHQ judge for years two and three.

This book tells of Cannon's journey along the road to TQM and beyond. Those builders who are willing to admit there is always a better way of doing things will find Gil's book fascinating and rewarding. It is a must-read for both large and small builders who are concerned about competing into the twenty-first century. But it is especially valuable to the small builder who has limited resources for undertaking a quality improvement program.

—E. Lee Fisher
Senior Industrial Engineer
NAHB Research Center

About the Authors

Gilbert J. Veconi and his wife, Barbara, have owned land development and small-volume home building companies in the Rochester, New York, area for the past twenty years. They currently build custom homes in the $200,000 to $700,000 price range in various Rochester suburbs and were winners of a 1993 National Housing Quality Award. Gil was a judge for the 1994 and 1995 NHQ Awards. He has also been both a panelist and speaker at several NAHB and management conferences on how to implement a quality culture in a home building company. Before starting his own company, Gil worked for Alcoa, Armco Steel, Alside Homes Corporation, and the former Ryan Homes in Pittsburgh. Gil holds a bachelor of science degree in architecture from Kansas State University and has completed graduate work in business and economics at the University of Dayton.

Charles A. Layne has over twenty years' experience as a college professor, academic dean, author, and management consultant. He has consulted with a wide range of both public- and private-sector organizations in the U.S. and abroad on quality improvement, quality assurance, and staff development. Chuck currently resides with his wife in the Richmond, Virginia, area. He holds a bachelor of science and a doctorate from The Ohio State University. Chuck and his wife had a new home built by Cannon Development, Inc., in 1985. Since that time, Chuck and Gil have worked together on TQM for Cannon and remain close friends (and "golf buddies").

Acknowledgments

First, I would like to acknowledge the many contributions made by the people who work at and with Cannon Development:

Barb Veconi not only kept me focused on this book but also on my commitment to TQM. Doug Allen, our company's production manager, was instrumental in getting our TQM effort off the ground. His ability to document the home building process gives us a sound basis on which to make the many construction decisions we face each day. Doug's undaunted commitment to continuous improvement has been and continues to this day to be the true strength behind Cannon's progress in satisfying customers. My son Craig, who is now a partner in the business, knew in 1988 that TQM was a better management philosophy. Craig was the first to bring TQM to my attention. Special thanks and acknowledgment also is extended to Ann Moffitt for all she does daily to implement our new way of thinking; to Rich Vallee and Todd Cymerman for carrying the TQM message to our suppliers and contractors; and to Sharon Parks for the many hours she spent transcribing my notes for this book. Finally, I want to say thanks to all the vendors and construction personnel who embraced the TQM philosophy with Cannon Development and who have worked tirelessly over the years to do things better for our home buyers. Cannon Development would not have won the NHQ Award without them.

I also would like to say thanks to Chuck Layne for introducing my son to TQM and for helping me shape my approach to TQM. He is as much a part of our success as anyone at Cannon Development. Without Chuck's practical expertise in TQM and his ability to teach others, we would not have achieved such success with TQM. I also could not have written this book without him.

Next, Chuck and I wish to thank the builders and other industry experts who contributed invaluable comments by reviewing outline or manuscript copy for the book:

Mr. Lee Fisher, NAHB Research Center, Upper Marlboro, Maryland, who also contributed the Foreword to the book; Ms. Barbara Von Bergen, J. A. Von Bergen Construction, Ltd., Alton, Illinois; Mr. Robert A. Bollier, Bob Bollier Homes, Inc., Leawood, Kansas; Mr. Martin Freedland, Organizational Development Associates, Inc., Atlanta, Georgia; Mr. Jeff Keil, J. Sargeant Reynolds Community College, Richmond, Virginia; Mr. Rick Martin, Renovations, Inc., Tulsa, Oklahoma; Mr. Fred Parker, Fred Parker Company, Inc., Fort Worth, Texas; Mr. Alan Simonini, Simonini Builders, Inc., Charlotte, North Carolina; and Mr. Peter G. Tibma, Nohl Crest Homes, Palm Harbor, Florida.

Finally, this book was written with the help and guidance of many people. Our spouses, Barb Veconi and Jan Layne, kept us focused and spent many hours with us sharing their thoughts, perspective, and "gentle" criticism on what we were trying to accomplish with this book. It is with much love, gratitude, and respect that we dedicate this book to Barb and Jan. We would also like to say thanks to Sharon Lamberton at NAHB for her guidance and patience. This book could not have been written without her support.

Publisher's Note

Destination: Quality was produced under the general direction of Kent Colton, NAHB Executive Vice President/CEO in association with NAHB staff members Jim DeLizia, Staff Vice President, Member and Association Relations; Adrienne Ash, Assistant Staff Vice President, Publishing and Information Services; Rosanne O'Connor, Director of Publications; Sharon Lamberton, Assistant Director of Publications and Project Editor; David Rhodes, Art Director; John Tuttle, Publications Editor; and Carolyn Kamara, Editorial Assistant.

Introduction

This book chronicles a home builder's successful journey into TQM. Before beginning the tale, let me explain why we at Cannon Development chose to make this journey and why I wanted to write this book.

In the late 1980s our company's financial position was sound. The new construction supply-demand balance was clearly in our favor and our net profits had more than doubled.

Year	Homes Sold	Net Profit (percent)
1986	25	7.1
1987	19	10.32
1988	23	14.74
1989	20	14.47

On the downside, however, we were making more mistakes. We had not made any significant improvements to the construction process. As a result we saw a marked increase in the total number of callbacks required for our new homes.

My wife, Barbara (who is co-owner and executive vice president of our company), and I were becoming increasingly uncomfortable with how we interpreted home buyer satisfaction. We believed quick responses to service calls could be a major factor. The way to increase buyer satisfaction, we thought, was to reduce the time it took to follow up on problems once the home was built. We were running fast to reduce response time but our customers still did not seem that satisfied.

By 1989 a timely response to callbacks had become a high priority. To protect our reputation for building a "quality" home, we continually pulled people off production to complete service work. This created problems for our contractors because they made little money on this type of work. Extensive service work also reduced the crews' production efficiency as they packed up their tools, drove from the construction site to an existing home, unpacked their tools, set up, performed the work, packed up their tools, drove back to the construction site, unpacked once again, and set up to work on the new home. Some of our contractors endured this time-consuming process more than once a week. Even more frustrating, the contractors sometimes were called back to correct situations that were not their fault to begin with.

Contractors grumbled openly that other builders they worked for didn't make them come back so often. Nor did other builders require as much service as we did. From what the contractors told me, it appeared that many of the other builders were telling customers the problems they reported "really weren't problems." Or, the builders claimed, the work was within warranty standards and therefore not the builders' responsibility to fix.

These callbacks had a second troubling aspect. We found it increasingly difficult to gain access to homes to do follow-up work because the vast majority of adults living in our homes worked outside the house. These owners wanted service work performed before they left for work in the morning, after they got home in the evening, or on the weekend. Owners sometimes forgot these appointments, and we would show up to find a locked house. This meant rescheduling and more time lost. Making arrangements to let us in was frustrating for us, but it was even more frustrating for the owners, who on several occasions reminded me that they had not contributed to the need for service to their home in the first place.

Barb and I also believed the social ties we maintained with buyers after the sale were a second indicator of our customers' satisfaction. We did not consider these social ties to be a critical measuring stick of customer satisfaction, but it was hard not to think that the numerous invitations we received to buyers' homes for cocktail parties, dinners, and holiday gatherings reflected a certain level of comfort and customer satisfaction. Nevertheless, I was haunted whenever the topic of conversation at these events turned from the usual chatting about kids and golf to the little but annoying problems owners might be having with their homes. When an owner who had hosted us socially called our office with a request for service, he or she often seemed apologetic or somewhat hesitant to be calling, as if concerned that our social relationship might be harmed because there was something wrong with their home.

Barb and I came to realize our positive social ties with owners reflected buyer satisfaction with the overall buying experience but not necessarily satisfaction with the home itself. We were doing a good job establishing strong personal ties during the sales and construction process, but we were still building homes with too many defects. Neither our buyers nor our contractors wanted to be inconvenienced by homes that were not built right to begin with.

I had a nagging fear that if we could not resolve the problems associated with increased service needs and improve our buyers' satisfaction levels, our reputation for quality would be in serious jeopardy. Our problems were clear, but our notions about how to solve them were not. *We needed help!*

In the summer of 1989 our situation began to improve. During that summer I attended a national conference on quality sponsored by the Rochester Institute of Technology. I was present to listen to my son Craig present the findings of his master's thesis research on contemporary perceptions of quality held by business leaders around the country.

Before the conference Craig had on several occasions offered to share with me some of the books on quality and customer satisfaction he was reading as part of his research. I had rejected even glancing at these books because I did not believe they were relevant to a custom home builder. Now, by coincidence, I found myself listening to speakers address the very issues of service and customer satisfaction that were rattling around in my head. What unfolded was a perspective on quality and customers quite different from the one I had held.

Until now, I had thought a quality home (i.e., a quality product and associated services) was defined by the grade of materials and sophistication of methods used in its construction. In this new approach, the definition of quality boiled down to producing a product (or service) that meets or exceeds customer expectations. The name given to this way of doing business was Total Quality Management, or TQM.

I noticed that the speakers emphasized that teamwork and communication among all those involved with producing a product (including the consumer) are crucial ingredients of TQM. However, they did not talk about how to control people's behavior. No one talked about how to crack the whip with employees or placate buyers. Instead, they presented ideas about how to control the process of building the product and how to be a better listener for buyers' needs and concerns. A business could cut costs and increase sales by increasing the satisfaction of (1) buyers (external customers) and (2) all those involved with sales, design, production, and service (internal customers). The proverbial light bulb was burning brightly in my head. The conference had challenged me to expand my perspective.

On some levels I still had a problem relating all this to my business. Just what did TQM really mean? Edward Deming's *Quality, Productivity and Competitive Position* (1982) and Philip Crosby's *Quality is Free* (1989) say TQM means meeting customer expectations. Deming seems to emphasize using statistics as a way to understand and improve work processes. Crosby, on the other hand, appears to say that TQM means improving management practices to reduce the cost of nonconformance to customer expectations. But something bothered me about both approaches. Home buyers don't think in terms of specific construction specifications. And many customers do not fully recognize their own expectations until they see the product in place or experience the service. How could I meet or exceed the expectations of my customers when they could not tell me what those expectations were?

The answer to my question would not come until several years after I had attended that quality conference. Over time, I have come to realize that I, the builder, serve as a bridge between customer expectations and standards. It is my job to determine what a home buyer's expectations are and then convert them to a set of standards. What this means is that our definition of TQM did not become fully clear until just recently. It's not that our thinking was *wrong* in those early days; rather, our thinking about TQM has evolved.

Initially we viewed TQM as both a business philosophy and a long-term business strategy. As a philosophy, TQM proposes that business success comes from satisfying internal customers and buyers. It is the responsibility of all who touch the work involved in building a home to achieve this end. As a business strategy, TQM means identifying customer expectations and examining the capability of the home building process to build homes that meet these expectations. This process includes all the actions and events from the initial sales contact through the end of the warranty period (and beyond, as we have found). Whenever the process is found incapable of meeting expectations, changes are made. The idea is to eliminate the need for rework.

People are the key to understanding TQM. The more folks are involved with identifying customer needs and improving the process used to meet those needs, the greater the odds for success. People who touch the work know more about that part of the process than those who don't touch the work. This expertise is invaluable when it comes time to "fix" a process.

I still think of TQM as both a philosophy and a strategy for running a business. More importantly, however, TQM at Cannon has come to mean an attitude that I share with each of our employees, suppliers, and contractors about how to build quality homes. Each of us wants to understand customer expectations and to do whatever it takes to see that these expectations are met.

I was cautioned early on that I *alone* could not manage the implementation of TQM at Cannon Development. And, there would be no knight on a white horse coming along to make things better. But if everyone worked together, we would have a chance.

Implementing TQM generally calls for a change in the way a company does business (in other words, in an organization's culture). Making the change requires lots of learning on everyone's part and is often described by Deming and others as a never-ending process. In many ways, what we must do to initiate and then develop this way of thinking resembles undertaking a long automobile journey—a metaphor we found useful in writing this book.

We chose to undertake our journey to TQM because its shift in focus from product grade to customer satisfaction made intuitive sense. If we succeeded in increasing both internal and external customer satisfaction, our reputation for quality would be secure. And our reputation was the basis of our continued success as a custom home builder.

Of course, home builders build homes; they don't, as a rule, write books. Why would I want to go through the pain and agony of doing something so different as writing this book?

While many business owners and executives extol the virtues of TQM, many others feel is it nothing more than an expensive fad. Through the story of how we implemented TQM I hope to convince you, as a home builder, as well as your contractors and suppliers, that you all can adopt the TQM approach without spending exorbitant amounts of money on consultants and training. We did it; so can you. As you will see, we did work closely with a quality consultant, Chuck Layne, who also assisted me with the writing of this book. For builders who wish to take a similar route we have included discussion of Chuck's evolving role (what it was and perhaps more importantly what it was not) and some tips on choosing a quality consultant. For builders who cannot or do not wish to work with an outside consultant, we have tried to provide in this story a good road map to help you get started, along with some detailed examples and a list of further resources.

Cannon Development won its National Housing Quality Award in 1993 largely because we were able to implement this new way of thinking about quality.* I feel a sense of obligation to share what we've learned with others. After listening to what others had to say, I was convinced TQM would be good for my business.

While we are a small-volume custom home builder, we believe what we learned carries a message for *all* those involved in build-

---

*The National Housing Quality Award is an annual competition co-sponsored by the NAHB Research Center and *Professional Builder* magazine.

ing homes. This book is my way to give back something to our industry. Indeed, as you read this book, should you feel more information about Cannon's experiences would be helpful in starting or continuing your own TQM journey, you are welcome to contact us through Home Builder Press.

Finally, I would like to become an advocate for the new home buyer. Many of the buyers I encounter work for companies that have incorporated some form of TQM. These buyers have extended the TQM concept into their everyday lives. This means they expect to be listened to. And many buyers, even now, are not listened to as they should be. We can make many effective adjustments to the home building process when we sincerely ask what our buyers want. Listen to what they say, and then do it.

Here, then, is our story.

Chapter One

Planning for the Journey

Like most home builders, I have known from the beginning that satisfied buyers are important to the success of my business. The question I have always struggled with, however, is how I could best ensure buyer satisfaction.

In 1989 I began to give serious consideration to the role Total Quality Management, or TQM, could play in achieving customer satisfaction. At that time, TQM proponents were adamant that the approach was appropriate for any American business. My gut reaction was that while this sounded good, I was still not sure it was right for us. And even if it was, I wondered, where should we start? Before committing myself and my company to a strategy that still seemed somewhat fuzzy in my own mind, I decided to seek some expert advice. I discussed my thoughts about quality and TQM for a home builder with a good friend of mine, Chuck Layne. A quality consultant who had experience assisting small companies implementing TQM, Chuck also was familiar with our organization and how we did business because he had purchased a new home from us in 1985.

After listening to me expound on my new concept of quality and my questions about the applicability of TQM to home building, Chuck suggested that before embarking on a journey into TQM, I should understand the economic climate in which we were doing business and how well Cannon Development was positioned to deal with economic threats and opportunities. Depending on what we

> ### Tip
> **The Role of a Quality Consultant**
>
> I wanted to work with a quality consultant who had experience helping small businesses *successfully* implement TQM. The consultant needed to not only understand the philosophical underpinnings of TQM and be a good teacher and mentor, but also provide evidence that TQM had actually been implemented in other companies similar in size to my own. Equally important, I wanted clear evidence as to how the consultant would "wean" himself from the company. I believe it is important to stand alone as quickly as possible. While not absolutely critical, it helps if the consultant has some familiarity with the home building industry. It was just coincidence that I had a friend who could help me. If you do not know of someone who can assist you with getting started in TQM, one option is to contact the NAHB Research Center for information on the TQM consulting services and training they now offer the residential construction industry. (See the resource list at the back of this book.)

learned from this analysis I could then decide whether or not to consider the TQM approach.

Assuming TQM was indeed appropriate for us, my next step would be to educate all Cannon employees about the TQM approach. If they were open to it, we would then be ready to take our first steps on the journey to TQM. What I heard Chuck telling me was to make sure my head was on straight before I committed our time and resources.

Was My Head On Straight?

At first blush, the task of describing Rochester's business climate seemed simple enough. But I soon realized this was a complex subject. For help in sorting everything out I turned to my wife, Barb, who is also Cannon's executive vice president and my chief advisor.

Together Barb and I looked for answers to Chuck's questions: What economic threats confronted our business? What opportunities could we take advantage of? What were our internal strengths (and weaknesses) related to these opportunities and threats?

After spending several hours pondering these questions, we realized we needed more information. Barb and I decided to talk with other people associated with our local home building industry.

We first spoke with our own employees and five or six of our most recent buyers. Next on our list were our real estate contacts, our commercial banker, a mortgage banker, two or three major distributors, three of our contractors, and members of the local home builders association.

I also spoke informally with five local builders I meet with regularly to discuss business matters of interest. Throughout these discussions, our aim was to identify and analyze dominant issues facing our business. We consciously avoided talking about possible solutions to make sure we first clearly understood the issues.

Combining our own observations with what we learned from talking with others, Barb and I identified some major trends confronting home builders in our area.

Rising Buyer Expectations

Conversations with buyers indicated that while customer expectations about the quality of our homes still tended to be lower than our ability to meet these expectations, this situation was changing.

The quality of our products and materials was not at issue. In fact they were the best available for the market we served. Our reputation for making things right when something went wrong also was positive. The issue was buyers' growing dissatisfaction with callbacks. Our readout of the situation was that while we were still ahead in the pursuit to please customers, the gap between our ability to build a quality home and the level of buyer satisfaction was shrinking. If this trend could not be reversed, the time would quickly arrive when the level of customer expectation would be higher than our ability to meet these expectations. I have illustrated this situation in Exhibit 1-1.

All of the people we talked with confirmed our initial observation that the level of customer expectations was on the rise. Moreover, these expectations involved more than a preference for minimal callbacks once the home was built. Buyers also were exhibiting less forgiveness for what we had accepted as the normal

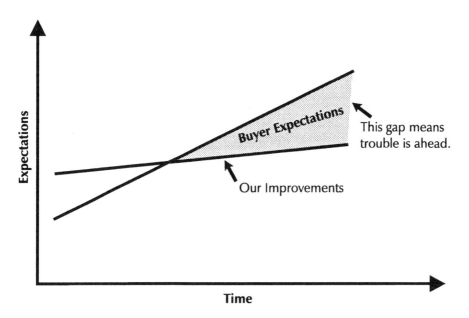

Exhibit 1-1: Buyer Expectations Versus Company Performance

complications of construction. I concluded that these increasing buyer expectations could eventually put our good reputation in jeopardy and put our competition at an advantage. If this occurred, our bright financial situation could turn sour.

Other Increased Competition

Everyone we spoke with in the late summer of 1989 also sensed an increase in competition. And this competition came not just from among other new home builders. Buyers who previously would have considered new homes were now giving more thought to purchasing existing homes or remodeling their current homes.

According to the real estate professionals with whom we spoke, many potential new home buyers were becoming wary of the perceived (and real) problems associated with new home construction. "Why be inconvenienced by construction delays?" or "Why not let someone else correct all the problems first?" were common reasons buyers turned to existing homes. To a lesser degree, the same arguments also were used to justify decisions to remodel. In combination these trends meant we faced a potential loss of market share. On top of that, we realized the number of home buyers (for both new and existing homes) was getting smaller.

Number of Home Buyers

By 1989 most of the "baby boomers" had reached their forties and were reaching their peak earning potential. As a result, the demand for custom homes was strong. Yet the builders in our area continued to experience fluctuation in demand. Buyers turning to the existing home market explained only part of this fluctuation. Further investigation revealed other factors.

Corporations with major installations in our area such as Kodak, Xerox, Bausch & Lomb, and General Motors were themselves feeling the pressures of foreign and domestic competition. Our local economy was at best stagnant and in some sectors had turned to decline. While I did not recognize it at the time, in 1989 our geographic region was heading into a serious economic slump. But I did recognize the local economy was not as strong as it once had been. The result was that fewer people were moving into our area. Historically, people relocating to the Rochester area represented 25 to 30 percent of our buyers. As relocation decreased, we stood to lose a significant share of our traditional market.

Our buyers' demographics also were changing. The number of empty nesters was on the rise and the average family size was getting smaller. More and more often, both spouses worked outside the home. At the same time, the number of people who planned to operate a small business in their home was on the rise. We could see that the needs of the new custom home buyer were changing. Still, we were unsure exactly how these changing needs would affect our business.

Cost of Doing Business

In 1989, when we surveyed our business environment, the cost and availability of building materials were relatively stable. But all builders, suppliers, and contractors were experiencing dramatic increases in the cost of workers' compensation insurance. For some time, we had been able to pass along these increases to the buyer. However, it was becoming painfully evident that the market would not let this continue. When higher insurance costs were coupled with the cost of rework and service, the cost of building a home was increasing too fast.

Our People

A review of the abilities and motivation of our employees revealed that they were all well qualified for their respective positions. More important, they all had positive attitudes about the business and were receptive to change. I also felt confident about the abilities of the more than fifty suppliers and contractors who worked with us. We felt that they all could provide a high level of workmanship and service.

On average, we lost only one or two contractors or suppliers each year. When we lost someone, it was because a contractor quit the business or left the area or because a supplier changed its focus from residential to commercial building.

However, I did discover problems with our contractors' labor pool. While the existing pool was able to meet our current demands for production and quality of workmanship, it did not have sufficient reserves to handle these demands should construction pick up. During my eighteen-plus years in building homes, I had watched the numbers of skilled workers decrease during economic downturns—and then be slow to rebuild when activity increased. In addition, several of our contractors were nearing retirement age without someone on hand to take over the business. We anticipated they would close down their businesses. If this happened, we would have to replace entire crews. Our ability to maintain a stable, qualified workforce was tenuous at best.

What Was the Problem?

I did not find anything in our analysis to suggest we were in the wrong business or that we were in the wrong niche. However, five trends indicated that we were headed for trouble if changes were not made to the way we approached building homes:

Trend 1: Increasing competition with existing home sales and remodeling.
Trend 2: Fluctuating pool of potential home buyers.
Trend 3: Increasing instability of a qualified work force.
Trend 4: Increasing level of buyer expectations.
Trend 5: Increasing cost of doing business.

The last two trends were particularly disturbing. I also was troubled by the decreases in productivity and increased costs associated with rework and service. Now that we were entering a soft market, I could not continue to raise the price of our homes to compensate.

Could TQM Offer an Answer?

While TQM will help a company take advantage of most business opportunities, it will not reduce threats to the business. Unless the threats are immediate and severe, however, TQM will help minimize their impact. A business facing impending financial doom must look to a business strategy other than TQM in order to survive. But a business that is basically healthy can help safeguard its health by implementing a TQM approach.

Although real economic threats loomed, our business had a strong balance sheet, well-qualified employees and tradespeople, and a good business reputation. We also felt we had some time to respond to the threat we perceived. Our conclusion was that Cannon Development seemed well positioned to implement TQM.

Confirmation and Consensus

Cannon employees knew I was considering TQM long before we officially brought it to the table. I did not do the groundwork described thus far in a vacuum. I gave employees several books on TQM to read (a partial resource list appears at the end of this book). When it came time to share my findings and to discuss the appropriateness of TQM with employees, I asked Chuck to sit in on the meeting. Chuck could call on his experience with TQM and other small companies to help explain key points and answer any questions that might come up.

The First Employee Meeting

I established three goals for the employee meeting. First, all employees would have the opportunity to critique my analysis. Second, the staff would be able to question my tentative conclusion that TQM was an appropriate strategy. Third, I hoped to build consensus on how contractors and suppliers should become involved with us. While we did not believe it was absolutely necessary for these groups to buy into TQM with us initially, we felt their buy-in would certainly be helpful.

At the meeting the general consensus was agreement on the accuracy and completeness of my findings. While our employees already had a basic awareness of what TQM meant, they were not sure it was appropriate for us or that it was our only alternative. Chuck was particularly helpful in explaining what TQM was and was not. By the end of the meeting, we had reached general agreement that we should give TQM a try. Cannon employees would have some training so that they could, as a team, better understand how TQM would function for us.

As for engaging the participation of our contractors, we decided our next step would be to see where they stood. I knew our suppliers and contractors would be interested in seeing what went into our decision to implement TQM. Some knew we were considering this but did not know the details. I anticipated they would

Tip

Employee Commitment

Employees may feel pressured to join in just to please the boss. I didn't want "yes" people, nor did I want "nay-sayers." However, I couldn't expect our employees to become TQM evangelists (at least not right away). I simply wanted everyone to commit to giving TQM an honest try.

My motivation for this position was and remains straightforward. My time and employees' time will not be wasted when critiques of ideas are honest. Employees always are welcome to seek clarification of my ideas and they should be willing to give any idea a try once clarification is provided. It's OK to still have reservations as long as such doubts don't get in the way of trying something out.

want some clarification of our goals, to know what role each supplier or contractor could play, and what we saw as the next steps.

Including Contractors and Suppliers

I decided to propose a partnership with suppliers and contractors as a way to define our relationship. All suppliers and contractors would be invited, but not required, to attend the training sessions we would be holding for Cannon employees over the next few weeks. We would not ask them to make a commitment until this training was complete. They could participate on our committees and work with us to solve common problems; they could take what they had learned about TQM back to their own organizations and give it a try; they could do both; or they could do neither. Regardless of what the suppliers and contractors decided to do, we would all at least have a common understanding of what Cannon was attempting to do. The staff agreed with these plans.

A Note About the Agenda

At this point we did not have a specific game plan for implementing TQM. However, Chuck helped me create a tentative schedule of training and orientation activities and anticipated outcomes for the next few months. He and I then confirmed these plans with the staff, and I began to prepare for our meeting with suppliers and contractors.

We kept the meeting arrangements simple. I would serve as chair. Chuck would serve as the facilitator. Doug Allen, our production manager, agreed to take minutes. We reserved a room big enough to hold up to fifty people in the lower level of the local public library.

We gave serious attention to the agenda. I would welcome the participants and introduce Chuck. He would then give a brief history and explanation of the TQM movement as well as some advice on how the people at the meeting could become involved. Next, I would summarize Cannon's efforts to date, present our proposal for suppliers and contractors to partner with us, and identify our tentative action plan for the next three months. Chuck and I would then field questions from the audience. We wanted the meeting to take no more than two hours.

Chuck and I agreed it would be wise to avoid using quality jargon during the meeting. Jargon can cause confusion and may turn some people off (it certainly does me). We also wanted to make it clear to the group that our decision to give TQM a try was based on Cannon Development's own analysis of the current business climate rather than on the recommendation of a quality consultant. Chuck's role at the meeting would be to provide explanations about various aspects of TQM. I would provide the practical application.

What was the incentive for suppliers and contractors to show up? I was banking on their curiosity and Cannon Development's good reputation as a builder and user of their services. In addition, I suspected many shared my concerns about the local business climate and would be interested in what we had to say. We called the meeting "Quality Improvement in the Residential Construction Industry," and the invitations we mailed out clearly stated there would be no obligation for further participation with us. We hoped the open door we provided would be in itself ample incentive to attend.

Over forty people, representing twenty-two contractors and suppliers as well as Cannon Development, attended the session. The reaction was quite positive and everyone seemed in general agreement with the various items we presented. We knew we had struck a chord when each company in attendance agreed to send at least one representative to future training. Even though the contractors and suppliers made their commitment to TQM conditional on how they felt after the training, we at least had everyone's ear—and they all knew what we were up to. We were ready to proceed.

What Was Our Destination?

As with any business, our destination has always been "success." For eighteen years I had believed the best way to get us there was to maintain a reputation for building a quality home and adhere to a competitive structure for prices paid to contractors and suppliers as well as for prices charged to the buyer. However, I now saw our road map somewhat differently. Pricing was still a part of the picture. But I was now paying more attention to the cost of rework and service.

An even more subtle change was a shift from caring about my reputation for building homes with good quality materials and construction techniques to caring about my reputation for building homes that please buyers (see Exhibit 1-2).

It was important that everyone involved in our journey understand this road map. The tool companies use most often to identify their road map in strategic planning and quality improvement activities is the mission statement. I had seen such statements posted in various corporate offices around town, but had not given much thought to developing one for our company. I now saw how creating such a statement could help for Cannon Development.

Before taking my first crack at writing a mission statement, I needed to do some more homework. Chuck gave me samples of corporate mission statements to review. I used these documents to develop a list of major points our statement should address. He also

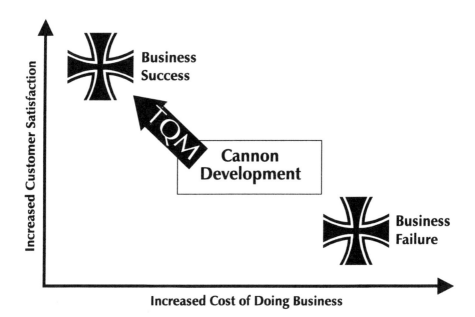

Exhibit 1-2: Our Road Map to Business Success

loaned me a copy of a book written by Thomas Falsey, entitled *Corporate Philosophies and Mission Statements*. (Additional resources you may find helpful in writing mission statements appear at the back of this book.)

From what I read about mission statements, I felt our statement should address three areas. First, I tried to clearly state our *belief* in customer satisfaction, improved construction processes, and people as key ingredients of business success. Second, I tried to identify the *principles* we would follow to satisfy customers, improve our processes, and involve people. These principles would reflect all the business policies and practices behind our approach. Third, I wanted the statement to serve as a *public affirmation* of Cannon Development's commitment to being a good corporate citizen of our community.

Because our mission statement would serve as a public statement of our business philosophy and operating principles, I wanted our employees, suppliers, and contractors to have a good deal of information about all the elements of our quality mission. I fully expected that the mission statement would be used as a gentle reminder by these folks should I ever waiver in the journey. I also thought our customers might like this information, although I didn't believe they would ask for the same level of detail.

It took me about one week to write the first draft of the statement. I then shared a copy with our employees and asked for their critique. I then made necessary changes and show the staff what I had done. We repeated this procedure several times over a period of about two weeks.

The process of writing a mission statement was enlightening. The exercise generated a good deal of discussion about what we were trying to do, how this could be clearly communicated, and our different perceptions of quality and working with people. Writing the mission statement helped us to create a road map for our journey and, for the first time, we now had common agreement on what all of us stood for as a company.

Our final mission statement (Exhibit 1-3) was fairly long. We use only the first section of the statement with buyers and the general public. It is reprinted, for example, on the back of our busi-

> ### Exhibit 1-3: Cannon Development, Inc., Mission Statement
>
> Cannon Development's mission is to provide the highest quality homes to its customers.
>
> Quality means using those products and procedures which provide our customers the best value with the highest degree of workmanship, along with expeditiously supplied services. We achieve quality by continually monitoring the needs of our customers and improving our processes and services.
>
> ### Guiding Principles
>
> The owners and employees will lead this long-term effort by subscribing to and daily implementing the following nine principles:
>
> 1. Create consistency of purpose toward improvement of our homes and services, with a plan to improve our competitive position and stay in business.
> 2. We will no longer live with commonly accepted levels of delays, mistakes, defective materials and defective workmanship.
> 3. Eliminate dependence on mass inspection. Instead, apply statistical evidence that quality has been built in.
> 4. Do not award business on the basis of price. Instead, depend on meaningful measures of quality along with price.
> 5. Work continually on improving the process of construction and service.
> 6. Institute training and be willing to be trained. Help people to improve and look to improve ourselves.
> 7. Eliminate fear so that everyone may work effectively towards this mission.
> 8. Break down barriers between contractors, suppliers, customers, and Cannon employees that prevent taking pride in what we do.
> 9. Replace work standards that prescribe numerical quotas with standards based on acceptable quality.

ness cards and on a plaque that hangs in our office reception area. However, we will share the entire document with anyone who wishes to read it—which occasionally does happen.

Final Preparations

We now understood why we were venturing into TQM, where we were headed, and in a general way, how we would try to get there. It had taken us approximately three months to reach this point.

However, we had some unfinished business to attend to before we could say we had truly begun our journey.

It also was time to formalize our working relationship with Chuck. He had volunteered his time to this point and I didn't think it was fair for me to continue imposing on our friendship. We also needed to decide who would lead us on this journey. And it was time to provide the training I had promised to all those going with us.

Working with a Quality Consultant

Looking back, I realize Chuck pushed us to do as much by ourselves as possible. There are things quality consultants should do for a company and there are things they should not do. Chuck did not assess our competitive and local economic conditions for us, nor did he tell us TQM was the appropriate strategy we should follow. He did advise us to investigate our business climate and then decide for ourselves if the TQM strategy was an appropriate match. He explained why a mission statement was important and offered tips on how one could be composed. *I* had to *write* the statement, although I must confess I would not have been able to do so without his guidance. Chuck could not tell us specifics about the roadblocks and detours we would encounter on the way to completing this mission. But he was able to show us how to identify and resolve such conflicts.

Chuck was truly my mentor during this time. I also made sure he was available to answer questions for any of our employees, contractors, and suppliers. He was also invaluable as a facilitator during the initial meeting we held with contractors and suppliers.

When we decided to continue working with Chuck on a more formal basis, he suggested I contact the regional New York Economic Development Office for advice on how to secure funds to cover his consulting fees and expenses. The department head informed me that funds were available to cover the costs associated with training for small businesses. However, they could not pay for consulting and facilitating meetings—which I have since learned is typical for most states. Most state agencies will only pay for training.

Chuck put together an outline of training we would need and an estimate of the expenses to produce a participants' manual for the sessions. It was up to me to submit this information to the state agency for approval. The agency then issued a contract to Chuck to cover his expenses for designing and duplicating the training materials as well as for his fee to train us on quality improvement techniques.

Leaders of the Expedition

An expectation I set for myself, Barb, and Doug Allen (our production manager) was to lead our quality effort by example. When a problem arose we would focus on causes and solutions rather than laying blame or pointing fingers. We would then ask those involved with the problem to help us solve it quickly. Daily business decisions were to be based on data and information rather than gut feelings and assumptions.

As leaders of the company and our TQM effort, Doug, Barb, and I were to "walk the talk." By setting this example we hoped others would adopt the same approach. However, I have to admit that none of us were particularly good at this in the early days. So we decided to set up a special committee to keep us on track and to provide strong leadership early.

We called this special group the Quality Steering Committee. Initial members of the committee were myself, as company president; Barb Veconi, Cannon executive vice president; Doug Allen, Cannon production manager; Ann Moffitt, Cannon bookkeeper; Ray Nichols, an independent painting contractor; and Chris Monheim, an independent concrete block mason. The group, chaired by Doug, planned to meet monthly once we were underway. Eventually some members would rotate—but consistency would be important at the start.

The specific charge to the Quality Steering Committee was to provide direction and guidance to our journey. The committee's charter is shown in Exhibit 1-4.

The role of this committee has never included working on specific problems. Rather, it is this group's responsibility to identify problems and then commission a group of employees and trades-

Exhibit 1-4: Quality Steering Committee Charter

1. Identify levels of buyer and internal customer satisfaction as well as changes in customer needs.
2. Identify problems with our home building process from initial contact with buyers through completion of annual service.
3. Commission special task forces to find solutions, and respond appropriately to task force recommendations.
4. Troubleshoot organizational problems that may interfere with our plans.
5. Monitor training needs and commission appropriate learning activities.
6. Make adjustments to our goals and plans to meet changing business needs.
7. Establish and maintain appropriate communication channels to all those involved with (and impacted by) Cannon's TQM effort.

---

people to find a solution. We decided to call these special groups *ad hoc teams*. The steering committee is not the only group that identifies problems, however. Any Cannon employee, contractor, or supplier can bring problems to the committee or they can work on the problems themselves and bring a recommended solution to us for consideration. The major responsibility of the Quality Steering Committee is to ensure that all recommendations are responded to and implemented when appropriate. Later chapters will present examples of issues addressed and how well this procedure has worked for us.

Training for the Journey

To function at its best, the steering committee needed some leadership training. Other employees, contractors, and suppliers who wished to participate in our journey also would need training on how they could best help out. Chuck recommended we train the steering committee first. Once this training was completed, we could then invite others for training. By the time of this second round of training, our journey would already be underway.

Steering committee training consisted of two 4-hour workshops, scheduled one afternoon per week over two weeks. At our first workshop we would:

1. Review our planning efforts to date (including the company mission).

> ### Tip
> **Who Should Head the Steering Committee?**
> While I was willing to chair our steering committee, Chuck advised against this. Because I was owner of the company, he argued, it would be too easy for committee members to defer to me on important issues. I wanted the group—not me alone—to provide leadership and guidance to others working with us. Chuck's advice made sense to me and I'm happy I followed his advice. Our production manager, Doug Allen, has served in this capacity since we started TQM. As a side benefit, Doug has been able to use his role as chair to develop his leadership skills.

2. Explain the committee's role (using the charter as a guide).
3. Select the committee chair and recorder.
4. Set ground rules for future committee meetings and activities.

The ground rules we agreed to follow (and continue to follow today) were:

1. Always have a written agenda (and submit it to committee members ahead of time).
2. Keep minutes of meetings in writing (and give them to committee members before the next meeting).
3. Start and stop meetings on time, and meet for no more than $1\frac{1}{2}$ hours.
4. Do not allow interruptions from others outside the meeting.
5. Complete whatever assignment is accepted and participate fully in committee activities.

At our second session we would:

1. Set specific goals for our journey.
2. Identify the detours and roadblocks that would keep us from completing our mission.
3. Decide how the ad hoc teams should go about their work.
4. Identify what we expected from the teams in way of a final report.

As it turned out, the decisions we made during our initial training have been extremely helpful to us over the years. Here's an idea of the further planning we needed to start the journey.

Setting Objectives. Our long-term goal is to remain a viable business by pleasing customers (both internal and external) and by reducing construction costs. While recognizing it may take years for us to be 100 percent successful, The steering committee nevertheless wanted to identify some short-term milestones to aim for over the next two to three years. I suggested three objectives:

1. Reduce the time to complete the final walk-through punchlist to 28 days.
2. Reduce the time to complete annual warranty service to 28 days.
3. Hold the variance between estimated and actual construction costs at or below 2 percent.

Achieving the first two objectives would require that we (1) educate the buyer on what to expect (so that callbacks were for legitimate reasons) and (2) root out and eliminate the reasons why those callbacks were needed. Achieving the third objective would mean we were in fact controlling the cost of doing business.

Success in meeting these more specific goals would clearly indicate we were making progress, which in turn would help sustain our momentum in our drive toward quality. The committee concurred these were appropriate milestones.

Anticipating Detours and Roadblocks. Using a brainstorming procedure, members of the steering committee identified nineteen possible detours, or roadblocks, we would have to eliminate in order to achieve our short-term objectives. Included on the list were concerns with communication, scheduling, uneven quality of materials brought to the jobsite, slow service turnaround times, and morale. The committee felt that four or five items would be about all we could handle at the start. Using a procedure called the Nominal Group Technique (see *The Memory Jogger* in the resource list) we identified five issues that should be given top priority during the foreseeable future:

1. Lack of written labor and material specifications.
2. Excessive time to complete final inspection and warranty service.
3. Need each spring to replace exposed porches poured during the winter.

4. Too-short lead time for delivery of material to construction sites.
5. Widespread disagreement on who was responsible for construction site cleanup.

We believed that all five problems could be resolved in a relatively short period of time, but we were unsure of how much time to estimate for each one. How we resolved this uncertainty will be explained later.

Problem-Solving and Ground Rules for Ad Hoc Teams. We wanted any group (or individuals) working on problems to follow a logical sequence of activities. From his experience with other businesses, Chuck recommended the following problem-solving steps:

First: Make sure that each person fully understands the issue being investigated in its simplest terms—What is the problem?
Second: Identify all possible causes of the problem.
Third: Identify all possible solutions to the problem.
Fourth: Select the most appropriate solution.
Fifth: Describe how and by whom the solution is to be implemented and then monitored.

These five steps describe the general process we continue to follow whenever we engage in problem solving. I should make two additional comments about our problem-solving approach. First, the five steps do not always follow a linear pattern. In actuality, a group of people working on a problem usually moves back and forth between the various steps. It is important, however, to start by making sure the problem is clearly stated before doing anything else, and not to jump to conclusions without first analyzing possible causes. Second, groups (and individuals) often find various techniques such as flow charting, check sheets, and cause-and-effect diagrams helpful in each phase of problem solving. I will discuss how we trained people to use such techniques in the next chapter.

As for how the ad hoc teams would conduct meetings, the steering committee wanted teams to operate much as the committee did. While we felt it would be all right for an ad hoc team to have additional ground rules, we insisted on only these three:

1. Don't hold a meeting without an agenda.
2. Within two weeks of accepting an assignment, give a preliminary report (either written or verbal) to the steering committee that describes the problem from the team's perspective and the game plan the team expects to follow in its investigation, including an indication of how long the team thinks it will take. At this time the team also requests any specific resources that are needed.
3. The *steering committee* selects the team leader.

This third rule is important. From what Chuck told us, the team leader's ability to facilitate meetings and keep people on track would be critical. To operate effectively the leader would also need the respect of the other people serving on the team. The committee and I felt this role was too important to leave the decision to popularity or chance.

Final Reports from Ad Hoc Teams. Once an ad hoc team had completed its business, the steering committee wanted a one- to two-page written report describing the problem addressed; the team's recommendation; and how, when, and by whom the recommendations should be implemented and monitored. The team leader (whom we later came to call the team captain) was to deliver the report in person to the committee. We also wanted the presenter to verbally give the committee an idea of what the team considered in its deliberations.

We Were Off!

Less than four months after initiating our planning efforts, we were ready to embark into the world of TQM. Hopes of all Cannon employees were high and everyone seemed upbeat. What happened to us in the early years of our journey is described next.

Chapter Two

First Days Out

Home builders may become frustrated with TQM because they don't know how to get started. We avoided this problem by taking the time for thorough planning. But even people who know how to prepare for this journey can become discouraged early. The time and activity associated with planning can give you the sense of having started, but it is only when you start to implement your plans that you really find yourself moving down the road. Cannon Development would not have gotten past the implementation stage had we not realized the urgent need to produce tangible evidence that we were actually *making* quality improvements, not just talking about it.

Chuck and I felt the best strategy would be to get Cannon employees, contractors, and suppliers to work immediately on meaningful projects that were also simple enough to guarantee quick solutions. Nothing would breed success like success.

We were striving to instill a new way of thinking about how we built homes. It would not be enough to merely encourage everyone to get involved with problem solving—I had been doing that for years. However, folks in our partnership needed to learn more about how to work together and with this new entity called a Quality Steering Committee. The first leg of our journey was to provide partnership training. To make the training effective we wanted it to be "hands-on" and to focus on real issues facing the company.

> **Tip**
>
> **Getting a Fast Start**
> Asking ad hoc teams to work on real rather than hypothetical problems during training provides participants an opportunity to see direct application of TQM principles to their work. This also means you are already involved with true process improvement. Working on real problems gives you a leg up on getting started fast.

Training Our Partners

Chuck organized the partnership training into four 4-hour workshops. The sessions were held from 2:00 to 6:00 p.m. on four consecutive Thursdays at our offices, which at the time were located in a model home. Twenty-five contractors and suppliers, the five steering committee members, and the other five Cannon employees participated. As with the earlier training for the steering committee, these workshops dramatically shaped our later TQM efforts. I would like to describe in some detail what occurred.

Of the five issues the steering committee had identified earlier, four seemed likely to be quickly resolved:

1. Who should be responsible for construction site cleanup?
2. How should we solve the problems associated with pouring exposed concrete porches during the winter?
3. We lacked written labor and material specifications. What would it take to put them in writing?
4. What should be done in order to give more lead time for requesting delivery of materials to sites?

The fifth issue, related to reducing the time to complete final inspection and warranty service, was more involved. We decided it should be put on hold until we had some experience working on the other issues.

We felt participants would see the four questions as real-world issues. Many of those in attendance would have experienced some aspect of these problems on the job. We also believed these questions were manageable and estimated that approximately 8 hours of workshop time would be sufficient for ad hoc teams to thoroughly examine each question. Chuck accordingly planned two full workshops around these four questions.

Because there were four questions, there would be four ad hoc teams meeting during training. From among the people we knew would be attending, the steering committee identified four individuals we could ask to act as team captains. Doug Allen secured their agreement to serve in this capacity before they came to the first session.

The First Workshop

During Workshop One, Chuck helped us get organized using this agenda:

1. Review our planning efforts to date.
2. Explain the roles of the steering committee and ad hoc teams.
3. Illustrate techniques the teams could use while working through the five problem-solving steps the steering committee wanted them to follow:

   First: Make sure each person fully understands the problem.
   Second: Identify all possible causes of the problem.
   Third: Identify all possible solutions to the problem.
   Fourth: Select the most appropriate solution.
   Fifth: Describe how and by whom the solution is to be implemented and then monitored.

In essence, Chuck talked about how to collect, organize, and interpret data and information to solve problems. He illustrated a variety of specific techniques the teams could decide to use in completing their assignment. *The Memory Jogger* (see the resource list) served as a reference on the "why's and how's" of these techniques. Knowing the temptation is to run with the first viable solution rather than explore all possibilities, Chuck also emphasized the importance of spending ample time making sure everyone understood the questions asked, the potential causes of the issues, and *all* possible solutions before coming up with final recommendations.

> **Tip**
>
> **Watch Out for Quality Jargon**
>
> Chuck and I agreed at the outset to avoid using jargon commonly associated with TQM efforts. He and I felt this would be particularly important during training. A term such as "statistical process control" is easily misunderstood, and jargon may intimidate some participants. I did not want terminology and theory to be the issue in training. Rather, I wanted the application of TQM principles, regardless of what they were called, to be the focus. We all encounter jargon in virtually any area of business we seek to improve (not just in TQM). While it may take a little time to translate "their speak" into "your speak," doing so will help ensure your grasp of the fundamentals and smooth the communication process throughout your company.

Doug concluded the workshop by announcing the team captains and setting up the four ad hoc teams.

Over the next day or two, a few of the participants called to inform me they could not come to the next workshop. For one legitimate reason or the other, their work schedules would not allow them to continue. Each caller expressed interest in what we were doing, however, and asked to be kept informed of how things were going. I was not concerned, as we still had a large and representative number of contractors and suppliers in attendance.

The Second and Third Workshops

Workshops Two and Three were devoted entirely to working on the four questions. A steering committee member participated on one of the teams. Other steering committee members acted as observers but did not participate. Chuck, Doug, and I rotated from group to group answering questions and discussing specific questions and dynamics that arose within a particular team. Doing this gave Doug and me an opportunity to see how folks were working together and the effectiveness of our five-step problem-solving procedure. In some groups, participants did not at first fully understand the problem being addressed. Often some participants would be discussing solutions while others were still examining causes. And of course, some participants tended to expound on war stories associated with the issues at hand. By gently reminding the groups to stick to the five steps—one at a time and in sequence—we brought a sense of order to the proceedings.

Midway through Workshop Three it was clear that none of the teams would complete their work during the time we had scheduled. While this situation was not totally unexpected, we all had hoped the teams could complete their work within the time frame set aside for this purpose.

> **Tip**
>
> **Training Plans Should Be Flexible**
>
> Chuck and I had a general idea of what we wanted to accomplish in our partnership training. Not having a canned program made it very easy to make adjustments to plans as we went along. Having this flexibility meant we were able to keep the training on track and meaningful to participants. However, this does not mean we did not pay attention to many important details. For example, training was always held in a location away from distractions; it always occurred in a specific time frame, and we did not run over the time allotted. We made sure participants saw reasons/benefits for being there and we did not try to cover too many topics in too short a period of time.

After a quick conference with Chuck and the steering committee, we adjusted our plans for Workshop Four. The ad hoc teams would spend the first half of the last workshop preparing a progress report. They would also determine what additional resources and information would be needed to complete their assignment. The participants agreed that this seemed like a reasonable way to wrap up the training.

The Fourth Workshop

During Workshop Four each team captain gave a status report to the steering committee. This was done in the presence of all participants. The four captains indicated their teams were making progress and wanted to continue working on their respective problems over the next month or two. In each case, they also indicated that they needed no additional information at the present time but stated that they knew they could come to Doug or me should they need anything.

Before adjourning Doug briefed the teams about the ground rules the steering committee believed would help the groups as they continued to meet. The teams were encouraged to edit the list of rules to fit their specific situations. As it turned out, none found this necessary, as the list was fine for them as stated. Doug also provided the teams with details on what the steering committee expected in the final reports. The fourth training session adjourned with a promise from each team that final reports were forthcoming and with a promise from Doug that he would be in touch with each captain periodically to see how things were going.

The participants told us they enjoyed this hands-on experience. Later a few of the contractors' and suppliers' companies initiated their own forms of TQM, using what they had learned in our training as a guide. Most participants also indicated they would be willing to participate on other ad hoc teams should their services be needed. A few indicated a willingness to sit on the steering committee when one or more of the current members rotated off.

Follow-up to the Workshops

The ad hoc teams continued to meet every two to three weeks after the training. Meeting more often was impractical, and this schedule was frequent enough to encourage progress toward their goals.

Cannon Development took specific steps to support the ad hoc teams. Meeting space was provided in Cannon offices and our receptionist offered to type minutes and reports. Such support may not seem important, but we have found it reinforces Cannon's commitment to TQM and to all the people who volunteer their time and effort to help us improve customer satisfaction. The team members recognize and appreciate that commitment.

The steering committee also continued to meet. Committee members were responsible for answering questions raised by the ad hoc teams along the way and for responding to ad hoc team recommendations. The steering committee members took both responsibilities seriously. Setting the agenda for steering committee meetings and following up on issues that arose kept Doug busy.

> **Tip**
>
> **Integrating TQM into Daily Work**
>
> One of the things I did to ensure TQM became an integral part of daily life in the office was to ask all employees (including myself) to do TQM-related work during normal work hours. I did most of my planning and writing during the day. Doug worked on steering committee matters and Ann set up ad hoc team meetings (and worked on meeting minutes) during their normal work hours. By contrast, employees who are required to do TQM work at home could easily question what advantage they gain by taking away time from their families.

Initial Successes—and a Failure

The four ad hoc teams began their work in mid-November. By the end of the following January, three teams had completed their work. Their recommendations were presented and subsequently accepted by the steering committee. The team working on site cleanup had met with several contractors to get their input on who should be responsible for keeping sites clean and to make sure they all understood why this was an important consideration. The team submitted an extremely well-written "Site Cleanup Policy" to the committee. We were able to send it to all our contractors and suppliers with only a few minor changes in wording.

The winter concrete team was dealing with problems that sometimes arose when we poured concrete porches during the winter months. For a number of years, at least one of the exposed porches we had poured during the winter needed to be replaced in the spring. These replacements were both costly and disruptive, and the steps we had been taking to avoid this situation did not seem to be working. I expected the team to come up with changes to our current procedure—in other words, that we would still pour concrete, but in a different manner. Their recommendation, however, was not at all I expected. They suggested we use temporary wooden platforms until spring, at which time the porches could then be poured. I had been insisting that porches be poured in the winter if we were closing on the home at that time. But the team was unable to find any procedures that would eliminate the problem. I had to admit their recommendation made sense, and the steering committee concurred. To this day, if we have a winter closing, exposed porches are not poured until spring! This has not caused any problems with closing. Very little additional money needs to be held back, since we are already setting money aside to cover sidewalk installation. And by educating the buyer about the procedure and why we follow it during the winter months, we don't have any buyer complaints.

The third team designed a format for written specifications and then wrote a set for wood siding material and labor. At this point in our corporate history these specifications already existed, but they existed in everyone's head. It was a relief to finally see them in writing. We would all now be operating on the same playing field when it came to siding! In addition, the format this team developed soon became the guide for writing all our material and labor specifications.

The fourth team, which was working on lead times for delivering materials to building sites, disbanded in early January. The team captain had been unable to find meeting times that fit everyone's schedule. As a result, the first two meetings scheduled after training were poorly attended. Two or three members out of seven did not provide the critical mass needed for the team to get anything done. There was too much investigation for such a small number of people to handle and the "horsepower" gained by having several members actively participating was also missing. The

---

Tip

Don't Let Expectations Get in Your Way

I have learned to watch out for prior expectations I may have regarding solutions ad hoc teams may come up with. Otherwise, I will have a closed mind to what the teams are trying to tell me. What I try to do is close my mind to such expectations so that I remain open to the best solution offered. It's amazing what one can learn by doing this. I still have lots of hopes and wishes about these outcomes; but by restraining my expectations I now am seldom disappointed.

members who had been attending regularly asked Doug if they could be reassigned to one of the other teams.

Lessons Learned

Watching the ad hoc teams and steering committee at work provided several lessons on how to do things right in the TQM process. We also learned some things not to do.

Training and Support Must be Immediately Applicable. On the positive side, I realized the importance of providing training to people in a way that allows them to immediately apply what they learn. The steering committee also had the opportunity to see how important it was to provide support and direction to ad hoc teams and then respond to team recommendations in a timely manner. Chuck had alerted us earlier about the importance of this support. Hassles with meeting arrangement, word processing, duplicating, and other support functions will dampen team enthusiasm. But nothing discourages a team more than lack of a timely and decisive response to their recommendations. We don't have to say "yes" every time; but we do have to explain our responses and do so as quickly as possible. We now saw how applying these principles had positively affected the day-to-day workings of our ad hoc teams.

In turn, the ad hoc teams were able to use their workshop training immediately. This experience reinforced the benefits of using teams to solve problems and the ability to approach the problems in a systematic fashion.

Quality Improvement Requires Time. I learned to be patient and allow the ad hoc teams do their work in their own time. It takes more time for a team to solve a problem than it would for Doug or me to come up with a solution. But the quality of team output and the positive impact using teams has on getting many people working together and involved with TQM more than outweigh the wait for a team to find answers.

The steering committee and I also learned that it is important not to underestimate the time it will take to resolve problems. But while we realized the significance of this, none of us were sure what we should do about time estimates. One team had already disbanded because time was a problem. We believed future teams would ask about this up front. And, we wanted to be in a position to give the team some idea of how much time they would probably need before we asked them to commit to an assignment. As it has turned out, however, most ad hoc teams are pretty good at estimating how long it will take to complete an assignment. The steering committee generally has not had to worry about coming up with an estimate.

Composition of Ad Hoc Teams. The steering committee came to two additional conclusions about how future teams should be organized. First, we noted that communication seemed to go more smoothly when a member of the steering committee sat on a team. We decided to make it a standing practice for a committee member—other than Doug or me—to sit on future teams. The committee felt our presence might put a damper on an ad hoc team's ability to have open and honest dialogue. Alternatively, our presence might become a crutch. Team members might feel they could rely on us for the answers rather than putting their own efforts into the investigation. Even worse, the team might feel we were there to make sure our wishes were obeyed. Finally, other teams might fear that priority would always be given to implementing recommendations from the "bosses' teams." While Doug and I would have no intent to dampen or distort team efforts, we both agreed the most effective way to prevent such problems was to not have us participate on ad hoc teams.

Second, we concluded that smaller ad hoc teams would be more practical. The seven- to eight-member teams organized during training had seemed manageable. However, attendance at meetings dropped once training was over. Typically the same five or six members would show up for each team meeting. The three teams that completed their assignments found that the absence of the additional members did not hamper their work. Also, two team captains expressed that finding convenient meeting times for five people was more realistic than for eight. The team that disbanded, on the other hand, did so because they found having only three members present at meetings was not sufficient to get anything accomplished.

After analyzing this situation, the steering committee decided future ad hoc teams should have five to six members:

> **Tip**
>
> **Diversity of Ad Hoc Team Members**
>
> We want people who serve on ad hoc teams to have a great deal of knowledge and experience with the issue under investigation. But just as I have to guard against making prior assumptions about solutions, so do these experts. That's hard to do. Our steering committee has found an effective control for this: We always select one member who has little or no experience with the topic at hand. This person can see the "forest *and* the trees" if you will, and is not afraid to ask questions other group members might avoid or overlook. Including a team member who is not a specialist will help keep the group from acting on assumptions or biases.

- One team captain (selected by the steering committee).
- One representative of the steering committee (other than Doug or me).
- One member who knows very little about the issue investigated (this would be the person to question everything and to help the team see the problem in perspective).
- Two to three people with expertise in the issue investigated.

The Diminishing Role of the Quality Consultant. Chuck had been instrumental in designing and facilitating the training we had offered so far. He advised Doug and me on how to assess how well the teams were able to organize themselves, run meetings, and apply systematic problem-solving principles. The three of us observed the teams in operation during the second and third workshops and took corrective action when warranted. Overall, the teams had performed quite well.

The steering committee also followed Chuck's advice on how we should assess final reports. Our original plans had called for final reports during the fourth workshop. All four teams had needed more time. I was somewhat concerned, as I had wanted Chuck to be present to observe these presentations and any ensuing deliberations by the committee (we didn't expect to make final decisions on the spot). However, Chuck suggested to the steering committee that this would be a good time to see if we were ready to go it alone. Training had gone well and everyone seemed comfortable with how things were progressing. This idea made sense to the committee. We were able to handle what happened during the team presentations without outside help. Chuck's active role as a consultant was now over; but as I will explain later, we would later call on him again.

We Needed A Breather!

At this point we thought we were off to a quick start. Within ten weeks, ad hoc teams had made three significant improvements to the quality of our construction process. Steering committee members were working well together and shared a common sense of purpose. We all had learned a great deal about how TQM should work at Cannon Development.

But the steering committee was beginning to grow weary. For more than four months, we had devoted several hours weekly to planning, training, and kicking off our TQM effort. Doug and I discussed our observations and we decided to give the group a breather. We did not want to lose the momentum that had built up, but we also didn't want to burn people out. Doug announced this plan at a committee meeting and the group offered a collective sigh of relief.

In hindsight, I realize Doug and I made a tactical error at this point: We neglected to specify the duration of the breather. Indeed, this error came close to bringing our journey to an end.

Doug and I used this "down time" to implement the three new procedures recommended by the ad hoc teams. It was midwinter in Rochester and we were scheduled to pour two porch floors soon. We wanted to see if what had been written on paper would work in practice. We also wanted to prepare our new site cleanup policy for mailing. And, inspired by the team's model specifications for wood siding material and labor, I decided it was time we created written specifications for all materials and labor. The steering committee had already identified written specifications as something we should have and the siding specifications ad hoc team had provided an excellent example to follow. So, while the committee was taking a break, I asked Doug to work with our two site supervisors to pull together a materials and labor specifications manual.

Developing Written Material and Labor Specifications

It may not be readily apparent why written specifications are so important to companies involved with TQM. The answer lies in understanding the relationship between customer expectations and specifications. Let me try to explain.

According to TQM theory, buyer satisfaction stems from the customers realizing their expectations have been met or exceeded. However, most buyers don't clearly know what their expectations are; and if they can express them at all, it is usually done in a very general way. For example, our buyers often select ceramic tile for various locations in their new home. They expect a tile floor to be attractive and functional. But customers usually have no idea of what goes into laying the tile or what is required to handle or maintain the material. In addition, they assume there will be no problems with the floor once it is laid.

As the builder, I am responsible for knowing both how to lay a tile floor properly and how to correct any problems that may occur later. As a responsible business owner, once I have determined the cause of any problems, I need to create or modify appropriate material and labor specifications to ensure the problem doesn't recur on future jobs.

When we talk about improving construction processes in order to promote customer satisfaction, we really are talking about improving material and labor specifications. But buyers don't think in terms of specifications. They think in terms of desires, wants, wishes, and performance. Builders, contractors, and suppliers think in terms of specifications. My role is to bridge this gap. In later chapters, I will explain how we at Cannon Development create this bridge. At this point, however, I felt it was important to create a written set of specifications to assist us with process improvement. Doing this would help us satisfy buyers *and* our internal customers—our suppliers, contractors and employees.

I asked Doug to organize the manual around the fifty different phases of our construction process. The numbering scheme shown in Exhibit 2-1 had been previously developed for estimating and

Exhibit 2-1: Table of Contents, Specifications Manual

502 Permits	551 Plumbing
504 Surveying	552 Rough Electric
505 Excavation	556 Garage Doors
509 Utility Service	558 Asphalt
510 Select Materials	560 Insulation
511 Footer Concrete	562 Drywall
512 Basement Gravel	563 Closet Shelving
514 Concrete Block Labor	564 Trim Material
516 Tarring	565 Trim Labor
517 Rough Grading	566 Shutters
518 Driveway Stone	567 Wood Deck
520 Conductors	568 Paint
522 Steel	570 Gutters
523 Rough Lumber Material	571 Ceramic Tile
524 Sliding Glass Doors	572 Cabinetry
525 Windows	574 Security System
527 Skylights	576 Shower Doors
531 Framing Labor	579 Resilient and Carpeting
534 Roofing	580 Hardwood
535 Siding Material	583 Mirrors and Glass
536 Siding Labor	584 Appliances
544 Brick Labor	586 Window Cleaning
547 Flat Concrete Material	587 House Cleaning
548 Flat Concrete Labor	590 Fine Grade
550 Heating	592 Landscaping

We use the same phase numbers for accounting and estimating. Leaving strategic gaps in the numbers allows additional specifications to be added in the future.

---

accounting purposes. The format of each specification would be the same as the one developed earlier by the ad hoc siding specifications team (see Exhibits 2-2 and 2-3).

Doug assigned the development of draft specs for various phases to the two site supervisors. He also assigned several to himself. Doug and the site supervisors were to talk with anyone whose work affected the output produced during each phase of construction. Everyone Doug and the supervisors spoke to was very coop-

Exhibit 2-2: Wood Siding Material Specifications

1. Wood siding material to be as specified on production plan to include change orders.
2. **Siding**
   - ½x6 Beveled: AYE grade western red cedar, kiln dried (19% or less moisture content), smooth one side, rough one side, smooth butt edge w/flat grain face. Not more than 10% of total lineal footage shipped will be in lengths 8 ft. or less.
   - ¾x8 Beveled: 95% select −5% quality (95%–5%) Western Wood Products Association (WWPA), solid tight knots (STK). Western red cedar graded rough one side and butt edge, partially air dried. Butt to be full 3/4 in. Shipped in 8 to 20 foot lengths. Not more than 10% of total lineal footage shipped will be 8 foot lengths.
   - 10 in. Channel Rustic: 95%–5% WWPA STK partially air dried western red cedar, graded rough one side, 8 ft. to 20 ft. lengths. Not more than 10% of total lineal footage shipped will be 8 foot lengths.
3. **Exterior Trim**
   - Rough Sawn Cedar: 7/8 in. nominal thickness inland red cedar #3 grade and better graded rough sawn one face and two edge, kiln dried.
   - Primed Finger Jointed Cedar: No knots, oil based primed four sides white or off-white. Smooth 4 sides and eased 2 edges. 1x2, 1x3, and 1x5 may be cut from wider boards with or without eased 2 edges.
4. **Air Retardant** 11# asphaltic felt paper 36 in. wide, 400 sq. ft. per roll or if shown in specs Tyvek or Equal Housewrap.
5. **Nails**

Material	Nails*
½ x 6 cedar	2 in. - 3/16 in. Head
¾ x 8 cedar, ext. trim and 10 in. channel rustic	2¼ - 7/32 in. Head

    *Double dipped galvanized "splitless" ring shank wood siding nail by MAZE.
6. **Sealant** Non sag siliconized acrylic, 11 oz cartridge Geocel 920.
7. All materials to be hand unloaded just off the house side of driveway.
   If site is muddy, materials to be stacked so ends or centers of boards do not rest in mud. Materials to be covered in polyethylene and held down with boulders/bricks when requested by builder or on judgment of driver based on weather conditions. Caulk to be left at builder's office if night temperature will be below 35 degrees. Materials to be returned to be stacked alongside driveway and covered. Based on weather conditions, materials to be picked up within 48 hours of being called (credits to be issued on next billing.)
8. Only builder may vary these specifications.

erative. All seemed to recognize this as a significant project in our TQM effort and they were happy to help out.

This project turned out to be quite a chore, mainly because there was so much to do and there were so many details to cover. Doug and I initially thought it would take six months or so to wrap things up. In the end, it took us a little over one year to describe 60 percent of the phases and another year to complete the rest. But complete them we did!

As portions of the specifications were completed we began to use them in our operations so we did not have to wait for the specs to be completed to benefit from the effort. To this day the manual is a valuable tool, not only for those involved with process improvement but also for educating new contractors and suppliers.

Developing a specifications manual was *not* a detour in our journey. It was a leg of the journey we had to take. But because this project was taking Doug and me so long to complete—and because it was occupying so much of our time—the steering committee began to feel left out of the process. While it would be some time before Doug or I fully realized what was actually happening, this issue was raised in the first "post-breather" meeting Doug called five months after the initial ad hoc teams had made their final reports.

Doug called the meeting to discuss some ideas he had picked up at a local seminar (sponsored by the Rochester Home Builders Association) on how to track the cost of implementing a quality assurance effort. Chuck was also present to review how other companies tracked these costs. The upshot of these discussions was to put the idea on hold. However, near the end of the meeting, three of the committee members indicated that while they were not turned off to TQM they most certainly had lost some of their momentum and sense of direction. This was a clear warning sign, but we ran out of meeting time before we could come up with a plan to correct the situation. And in fact, Doug and I were still so focused on completing the specifications manual, we were not really paying attention to what was happening!

This issue hit me squarely between the eyes at our next steering committee meeting (four months after the meeting described above). The first item on the steering committee's agenda, which

Exhibit 2-3: Concrete Block Labor Specifications

1. Footers, walls, piers and cores to be per production plan to include change orders.
2. Mason is responsible for accuracy of foundation location to the extent of complying with offset, sideline, and setback dimensions as staked by the builder's engineer.
3. Footers to be dug (by excavator) 5 in. below cellar grade to solid ground or certified compacted fill. Stepping of footers where necessary to be in 8-in. increments. Builder to straw footer excavation when temperature will be below 32 degrees.
4. Mason to form both sides of all footers and all sides of piers (builder to supply and mason to maintain forms). Footer to be sleeved with 2 in. PVC (supplied by builder) by mason at locations as determined by builder. Builder to schedule town inspection for footer prior to pouring of concrete. Concrete to be ordered by builder per footer concrete phase specifications. Temperature to be 25 degrees or higher when pouring footer. Mason to straw footer after pouring when temperature will be below 32 degrees.
5. Builder to schedule delivery of pea gravel and block. Mason may elect to install one load of gravel after forming and before pouring footer. Mason may also elect to set first course of block in concrete after pouring footer.
6. Builder may elect to vary wall configuration to accommodate grade. Grade line to be established to builder prior to parging.
7. Mason to use Type S mortar, or 2 to 1 mortar portland mix. Builder to supply all material to include mason sand (mason is responsible for warming when necessary). Temperature to be 25 degrees or higher when laying block. Anti-freeze to be used in mix November 1 through March 31.
8. Builder to have water service available to mason, on site when possible.
9. Walls to be plumb, level and dimensionally correct within $1/4$ in. Diagonal dimensions to be correct within $1/2$ in.
10. Mason to leave weeps every third block to bottom course by omitting mortar from bottom half of the inside joint.
11. Water line to enter the house through the foundation wall between the 2d and 3d course of block.
12. Porch and garage walls to be tied to basement wall with 2 in. lengths of durawall installed every other course in the top 6 courses of block.
13. Fireplace foundations to include installation of cleanout door and thimble where applicable. Fireplace block to be run after framing when masonry wall will adjoin a framed wall.
14. All exterior joints above grade line and all interior joints above bottom of basement floor level to be pointed and struck.
15. Beam pockets to be properly sized and located, with cores of block to 3 courses directly below beam pocket filled with solid mortar.
16. Parging to be applied with minimum 1/8 in. thickness from footer or 4 in. below top of basement floor height at buried block to a height 2 in. below finished grade. Parging to be coved at footer.

appeared at the request of one of the members, was to decide whether the committee should continue. Doug and I do not remember who suggested this item. But the committee most certainly now had our full attention. If the committee decided to disband, we were sunk. Without their continued involvement, TQM would become just another good idea gone sour. I could not carry on alone.

The committee did not disband. Knowing this topic was on the agenda, Doug came prepared to explain what he and I had been doing over the past nine months with regards to the specifications manual. He also explained that he and I now both fully realized how other members were feeling and apologized for not taking corrective action earlier. All of the members indicated they accepted his explanation and apology. As a group, we then decided to refocus our efforts on what our internal and external customers might be telling us about needed quality improvement. Looking back, I realize this is the direction I should have encouraged the committee to take once they had caught their breath after the initial round of ad hoc team deliberations.

What Were Buyers Telling Us?

The steering committee recognized that most, if not all, of our efforts to date had focused on process improvement from the perspective of the internal customer—ourselves and our contractors and suppliers. We had not yet paid any significant attention to buyers as a group.

The Customer Satisfaction Survey

For several years, our company had sent a customer satisfaction survey to buyers six months after closing. When something negative came back on a survey, Doug or I would call the homeowner to make sure we understood the problem. If corrective action was warranted, we would then schedule the appropriate service. But to be honest, I was not aware of any trends or common complaints, since we had not taken time to collate the surveys.

The steering committee agreed that something should be done with the survey information. We might find some interesting things about what was bothering our customers. After a quick compilation of the surveys we had on file, a trend did emerge. The most common complaint from new homeowners was that cracks appeared in concrete basement floors sometime after closing.

The committee commissioned an ad hoc team to investigate the issue of cracks in concrete floors. Team members included one of our site supervisors and our concrete mason. We wanted at least one member who had little or no experience pouring basement floors. This person would not be hampered by prior assumptions about what causes cracks in basement floors. We also wanted someone from the steering committee to sit in. Both requirements were met by asking Ann Moffit, our office accountant, to participate. We also asked Ann to serve as team captain. She is a good group facilitator and would have the easiest time coordinating the schedules of other team members. We also secured the commitment of cement suppliers and our engineer to serve as special resource people to the team. While the team had only three members, another three to four individuals volunteered to work directly with them.

At its first meeting, the team could not come to grips with the issue. They were unsure of precisely what they were charged to do. Doug and I subsequently met with the team to discuss the issue in greater detail. We reviewed the surveys and explained what Cannon was looking for, but we did not suggest any possible solutions. The team now had a much clearer idea of the problem.

A Valuable Lesson

I had just learned a valuable lesson. The steering committee must make sure an ad hoc team is given as much written information as possible up front on the issue to be addressed. This includes a cover memo concisely describing the problem to be solved. The committee also pays attention to the status report submitted by an ad hoc team soon after accepting an assignment. This early status report tells us if the team has really understood its assignment.

The concrete floors team submitted a final report approximately six months after beginning its work. Overall, it was a good report. One recommendation was to educate the buyer on the causes and likelihood of cracks occurring in concrete basement floors. A tremendous amount of time and effort was spent analyzing why cracks appear in the first place and what could be done to keep them from happening. However, two other recommendations—(1) using rebars throughout and/or a special mesh at the corners and (2) scoring the floor—were potentially expensive. Doug and I were also unsure whether both recommendations were needed or whether just one would do.

The steering committee asked the team to refine the last two recommendations. Cost estimates indicated each procedure would cost approximately the same. The team also determined that if both procedures were used, the problem with cracks would most likely go away. However, scoring alone would most likely eliminate 80 percent or more of the problem. Ann resubmitted her team's report with this new information within a month.

The steering committee accepted the first recommendation concerning buyer education. From this point on, we would include information on concrete cracks in the educational material we provide buyers at closing. We also would discuss the situation with the buyer during the final walk-through. The committee deliberated for some time on what to do about the other two recommendations. Our final decision was to go with scoring. The team had felt confident that this procedure would eliminate the vast majority of problems. While the team also recommended using mesh and rebar, they were less sure about the real impact of this solution. Doug and I saw to it that appropriate material and labor specifications were modified and, as it has turned out, scoring the floors and educating buyers about concrete has eliminated the problem.

The question over cost brought an interesting side issue to the forefront. Steering committee members confessed to wondering for some time how and when I would say "no" to an ad hoc team. I had told the committee I would support their decision to accept recommendations if the recommendations were based on thorough investigation. That's why we clearly laid out our expectations for how an ad hoc team would approach an issue and what detail they would include in their final report. I had not specified, however, that I wanted to know how much each recommended change would cost. To put this issue to rest, the committee clarified its expectations for final reports from ad hoc teams. Henceforth, the cover letter to ad hoc teams included the instruction that their report must include any costs saved *or* incurred by adopting particular recommendations.

I promised to provide any ad hoc team, regardless of who served on the team, with financial information they would need to determine the cost of making a change to procedures. I also wanted any team to have a complete picture of how our financial situation—including profit margins, overhead allocations, and construction costs—were affected by the issue on which they were working. The only information I was unwilling to share was the salaries we paid to individuals. As I began to follow though on this promise, I was amazed at some of the comments from recipients of the information. Many told me they were surprised I would actually give them this information. It seems there was quite a mystique about a builder's costs and margins. That mystique no longer exists within our company and we have yet to experience any problems created by making this information available to the people involved with our TQM efforts.

Continuing the Journey

We wrapped up the work of the concrete cracks ad hoc team almost a year and a half after beginning our journey into TQM. Most of the issues identified at the start of the journey had now been put to rest. Employees, contractors, and suppliers had remained active on the steering committee and various ad hoc teams. In addition, almost all of our contractors and suppliers had worked with Doug and me to help write the specifications manual, which by now was almost complete. However, TQM requires a constant refreshing of one's awareness of internal and external customers' priorities. At this point we wondered if there was anything else on the minds of our suppliers and contractors.

The steering committee formed a new ad hoc team to look into what might be causing our contractors to spend unneeded time on the job. Were there ways, for example, we could help contractors save through reduced cycle times? The committee also wanted the team to investigate whether or not our contractors and suppliers were having problems with *us*. Two site supervisors, our office accountant Ann Moffit, and our finishing and masonry contractors agreed to participate on the team. Because Ann had done such a good job with her previous assignment, we asked her to serve once again as team captain.

The team divided the list of contractors and suppliers among themselves and proceeded to interview each one about problems they encountered while working on our jobs. It took the team approximately six months to complete their assignment. Their final report was a typed summary of the comments gleaned from the interviews.

Most of the contractors (and all of the suppliers) were able to cite by example a multitude of problems they had encountered. Taken one at a time, the problems often seemed minor. Collectively, however, they indicated a lack of respect for each others' work, a need for better scheduling of work by Cannon, and a need for some changes in our material and labor specifications.

The steering committee accepted the report and asked Doug and me to address the various issues. We approached the issue of lack of respect by talking individually with the contractors whenever one of us had reason to meet with them in the field. We kept each other informed of whom we had spoken to so as not to duplicate our efforts.

Doug and I tried to impress on each contractor that they should think of the group which follows them on the job as their customer. The responses were interesting. Most contractors said no one had ever raised this issue with them before (at least not in such an emotional way). Some said they were not aware the problem even existed. Everyone accepted our suggestion to be more considerate and the new attitude seemed to take hold quickly. We have remained proactive in this regard, however; Doug or I make it a regular practice to ask contractors if they have problems. Occasionally they do, but a quick conversation with the offending party is generally all either of us has to do to resolve the issue.

Doug and I also rethought how we handled scheduling. We decided to make more of an effort to be on site when time came to start a new job. Doing this has also helped us solve many of the problems we were having with short lead times for requesting material delivery. (This was the issue, you may recall, that one of our first ad hoc teams gave up on.)

The real lesson we learned from this experience was the importance of *listening to our internal customers*. They have a lot to tell us, and it generally requires little work to resolve the things that bother them. We just have to take the time to find out what those things are.

How Long Were Our "First Days Out?"

The "first days" of Cannon Development's journey into TQM lasted for over two years. However, they were a productive two years. Our decision to get ad hoc teams involved early with real quality improvement activities fueled interest in tackling other problems. Our initial successes created an appetite for more successes. Starting with solvable problems also helped forge strong relationships and fostered trust among our partners. With confidence in the process and each other now established, the steering committee and ad hoc teams were now positioned to tackle other, more difficult projects.

We made mistakes, and our TQM effort even flirted with failure during the first year. But we managed to revive the waning enthusiasm of the steering committee by refocusing our attention on what had worked for us early—listening to our customers. By the end of our second year, we were listening to both internal and external customers.

It is important to note that I did not dismiss any employee, contractor, or supplier who lost interest in TQM (and I haven't to this day). And in fact, with the exception of the steering committee (who were reacting to fatigue after the initial push), not that many people truly lost interest. I suspect this was because Doug and I were constantly talking about team successes and the specifications manual at every opportunity in the office, out on jobsites, or during the many telephone conversations he and I had with contractors and suppliers. We kept people informed, which continued to spark their interest.

Since the early days of my company, I had maintained a practice of meeting once or twice a year with all employees, contractors, and suppliers for the sole purpose of getting their collective input on how we could better operate in the field. In the past, good suggestions sometimes were ignored because we did not have an organized way to do anything about them. TQM now provided us with a mechanism for doing something about such ideas.

We were now well into our journey, but we wouldn't fully realize our progress until much later. Other companies may find themselves "on the road" without recognizing exactly when and where they made the transition from planning to traveling. The next three chapters describe in more detail what life is like during the journey towards quality.

Chapter Three

On the Road

Our ultimate goal as a builder is to satisfy both buyers and internal customers in the most cost effective way possible. For more than six years now, Cannon Development has used TQM as the path to this destination. Our "road map" calls for a focus on customers, continuous improvement to the extended process of building a home, and help by all those who work with us to make these improvements.

The focus of our daily work is on pleasing both internal and external customers. Prior to starting the TQM journey, however, our company's primary focus was on pleasing the buyer (our external customer). While I always felt Cannon Development employees, contractors, and suppliers were important, I just did not see them as "customers" and neither did they. Now we all do.

Cannon Development regularly asks all customers about their needs and expectations for the extended process we use to market, sell, build, and service new homes. Buyers tell us what they expect in a new home; internal customers tell us what they expect about the construction processes used to build the home. By comparing their expectations to the building practices we use, we learn *where* and *when* improvements should be made. The people who work on the particular process or practice in question are then asked to determine *how* to make necessary improvements.

This chapter examines what all this means to me as a home builder.

> **Tip**
>
> **TQM Is Not a "Program"**
>
> I first realized TQM represented the way we do business when I found myself no longer referring to our "TQM program" when describing to others what we were doing. The word "program" implies something that can be described in physical terms, something with a distinct beginning and ending. And I cannot think of a slogan or logo that would adequately communicate its meaning. A program also lends itself to slogans and special logos. Likewise, terms like "ad hoc teams," "statistical process control," and "mission statements" cannot adequately describe what TQM means to us. I believe TQM is an attitude, and I can only point to its beginning.

Converting Customer Expectations into Process Improvement

If I could use only one phrase to describe the key to TQM it would be "listening to customers." Listening is how we discover what buyers expect in their homes and what suppliers, contractors, and staff expect about the home building process. Listening also tells us a lot about when and how our home building process fails to meet these expectations. Customers are satisfied when their expectations are met. Customers are delighted when their expectations are exceeded.

The steering committee reviews feedback from both internal and external customers on potential process improvements. We act on what we learn from customers in two ways. If we should correct a situation and already know how to resolve it, the committee simply asks me to see that necessary action is taken. If we don't yet know of an appropriate solution, the committee establishes an ad hoc team to investigate the issue and recommend a course of action. Once a recommendation is accepted, the committee then charges the appropriate people with making the needed changes and monitoring the results. In both instances, the committee makes sure the people who raised the problem are kept informed of what happened to their suggestion.

We also "listen" to customers in a more informal way. Day-to-day interactions between Cannon Development, buyers, and internal customers often produce common-sense ideas for procedural changes that involve only a few individuals, are minor in scope, and typically do not call for a change to any of our documentation. It would be overkill to subject such issues to committee review and approval. In such instances, I want the individuals involved to take action on the spot.

There is no magic formula for what issues should be resolved in the field and when should come before the steering committee. We base our decision on the number of trades involved. If it's two (maximum three) we get them together on the job for resolution. More than this usually means we must come to agreement on the root problem, assess the impact of the proposed solution(s) and include cost-benefit analysis. This is better done through the steering committee.

Since the very first days of our journey I have openly encouraged people to accept the responsibility to act on their own when they felt it made sense to do so. I reinforce such behavior by applauding whenever I see it happen. In addition, the steering committee has agreed to consider only issues they feel warrant a formal response. When the committee believes the individuals involved should handle the situation themselves, the committee tells them so.

Our employees and other internal customers have received a good deal of training on individual and team problem solving. But more importantly, they have had many opportunities to practice what they have learned. I have taken every opportunity to reinforce the idea that TQM means each of us must *accept the responsibility to do what it takes* to satisfy all customers in the most cost-effective way possible.

Discerning Buyers' Expectations

We continue to listen to buyers throughout the extended process of building their homes. In fact, we continue to listen to what buyers have to say about home building well beyond our one-year warranty period on their new home. However, how we listen and what we learn varies depending on where we are in the process.

Listening During the Marketing Phase. Cannon Development uses focus groups and surveys to develop a general idea of what buyers are looking for in our offerings. Early in the planning of a new development, we'll place an ad in a couple of the local newspapers announcing our intentions and asking for volunteers to participate in the focus groups (see Exhibit 3-1). At the same time we mail a survey to potential new-home buyers interested in the area (and price range) in which we plan to build. A second survey is sent to real estate agents who work in the area. The survey sent to potential new home buyers is a slightly abbreviated version of the one used for consumer focus groups (shown later in Exhibit 3-2).

We make it clear in both ads and surveys that we are only interested in getting a general idea of what people are looking for in (1) a new home and (2) the community in which it is built. We do not make a sales pitch to participants at any time during or after meeting with us or completing a survey, although we do offer real estate agents who complete their surveys a $1,000 bonus for bringing us buyers of homes in the new development. Focus group participants are sent a letter confirming their participation, providing them with the time, date, and location of the group meeting (and directories to it), and thanking them for their interest.

The number of focus groups we form depends on the number of people who are willing and able to discuss what they look for in a community and a new home. Typically, we end up with two groups of six to eight participants. One group will be people interested in townhouses, a second will be people interested in detached homes.

Exhibit 3-1: Sample Newspaper Advertisement for Focus Group

[Name of Development]

Builder Seeks Participants for Focus Group.
Please share your thoughts for better new homes and architecture. We will reward your time and ideas.
Groups forming for homes and townhomes.
Homeowners/homebuyers in $120,000 to $170,000 range are particularly encouraged.

Cannon Development, Inc.
[Ad also gives contact telephone number.]

Each focus group meeting is held in an informal setting such as the local town hall and lasts less than 2 hours. Even when the locations are somewhat formal, we keep the meeting atmosphere informal. A person who does not work at Cannon Development facilitates each meeting. (We find local high school guidance counselors make good facilitators.) Someone from the company (usually my son Craig, who is now our sales director) meets with the facilitator before the meeting to review the agenda (Exhibit 3-2). Craig is not an active participant in the meetings, although he does greet participants as they arrive and introduces the facilitator.

Participants have an opportunity to discuss their ideas of an "ideal" community and some of the general design features they look for in a home. This gives us a general idea of what future buyers of our homes are looking for. For example, one of our new developments had plans for townhomes (something we had not tried before). Our designer suggested we incorporate some "flex space" into the floor plan of the townhouses. Such space could be a small family room off the kitchen, a formal dining space, a study, or a first-floor guest room.

The focus group for this development told us they liked the idea of flexible space in the floor plan. We also learned some things about their preferences regarding how this space might be used. Empty nesters tended to want a dining space or guest room, whereas young couples and singles would rather use the space on the first floor for a study or kitchen extension. However, we also learned some things about our design the participants did not like. The second feature suggested by our designer was to give the front elevation of the townhomes a "cottage look" by using window designs to take the focus off the garages (which typically are on the front of townhomes built in our area). While everyone seemed to like what we were trying to do, they did not like the "look." They preferred more traditional designs even though such designs did little to play down garage doors.

The group focusing on detached homes in the new development was likewise helpful to our planning efforts. They suggested the way in which we proposed to use street lighting, sidewalks, and landscaping seemed to give the neighborhood a small town feeling.

Exhibit 3-2: Focus Group Agenda and Survey

Outline for Focus Group Leaders

Facilitator Questions

Encourage participants to think in terms of what they have or haven't liked about their homes in the past, what they would like in a new home, how they have felt when purchasing or building homes.

I. Exterior architecture/landscape: 20 minutes.
   A. Elevations
      1. Review
         Which is most appealing?
         What about that plan makes it most appealing—use or looks?
         Examples of use: front porch, sidelights at front door.
         Examples of looks: brick, dormers, transom glass.
         What don't you like about each?
         Why? What would you add to any of these to make it better?
         Encourage participants to draw on plans provided, which we will collect.
      2. Color schemes and design
         Given the size of homesites (closeness of one home to the next)...
         a. Do you think it would be beneficial for the builder to limit the variety of exterior color choices, so that the neighborhood would look more uniform?
         b. Do you think it would be beneficial for the builder to limit the variety of exterior color choices, so that the neighborhood wouldn't have unsightly color combinations on some homes?
         c. Would the selection process be easier for you if you could pick a color-coordinated package of brick, shingles, siding and trim from 4 or 5 choices, as opposed to having to pick each item individually?
         d. Would you rather pick each color or material yourself?
         e. Would you be put off if you couldn't?
         f. Do you think that there are colors or materials that absolutely should not be used (e.g., green roof shingles, garage door screens)?

   B. Miscellaneous
      1. How important is it to have brick on the home?
      2. Is a wood deck substantially better than a patio?
      3. Is a flat driveway an absolute necessity?

continued on next page

II. Interior architecture (floorplans and the relationship of rooms): 20 minutes
   A. Is a formal dining room or a small family room worth paying more for?
   B. Is a cathedral ceiling in the living room worth paying more for?
   C. Are more windows worth paying more for in monthly utility bills? How much? Where would you want more windows?

III. The sales process (getting information): 15 minutes
   A. What information do you want to get from real estate salespeople? The builder's salespeople? (Examples: mortgage information, pricing) At what point in time?
   B. Do you want a price list right up front?
   C. The last time you shopped for a home, had you been prequalified for a loan amount? Would you want a representative of the builder to do this for you? How much help from the builder do you want in securing a loan?
   D. What are your greatest anxieties about buying a new home?
   E. Your biggest concerns?
   F. What do you expect from a new home salesperson?
   G. What do you not want in the sales process? What is too pushy?

IV. Features/specifications (allowances); includes survey: 20 minutes
   A. Are you comfortable when a builder gives an allowance for cabinetry, carpet, etc., or would you rather choose from a standard selection of materials and pay for upgrades?
   B. What is the fairest way for a builder to price upgrades or extras? What will you feel most comfortable with?

V. What else? 10 minutes
   A. What have you always wanted to tell a builder or architect?
   B. What is your biggest complaint or problem with homes you have owned?
   C. If you have never built a home before, why not? Is the building process intimidating? Why?
   D. If you have built before, what were the biggest problems? The biggest benefits?
   E. Is it important to you that your neighborhood have an entrance sign (to establish character/identity)?
   F. Should it be lighted? What about street lighting? Yes or No?

Survey for Focus Group Participants

Your age: _____

Please answer the following questions by marking the most appropriate box.

Your marital status:
  ☐ single, never married   ☐ divorced, with children
  ☐ married, without children ☐ divorced, without children
  ☐ married, with children   ☐ widow, widower

If you have children, what are their ages (check all that apply)?
  ☐ 6 years or under   ☐ 6 to 19 years   ☐ 19 years or older

Do you:
  ☐ own a home   ☐ rent a home   ☐ rent an apartment

What is your annual household income?
  ☐ $30,000 or less   ☐ $50,000 to $60,000
  ☐ $30,000 to $40,000   ☐ $60,000 to $70,000
  ☐ $40,000 to $50,000   ☐ greater than $70,000

Which of the following items would be a motivation for you to move (check all that apply)?
  ☐ greater security of home and family
  ☐ closer to job and/or shopping
  ☐ more attractive neighborhood
  ☐ better schools for children
  ☐ better job/higher salary
  ☐ more children in neighborhood
  ☐ growing family/need larger home
  ☐ want or need first floor master bedroom
  ☐ opportunity to own vs. rent
  ☐ want a smaller lot
  ☐ want a larger lot
  ☐ want less maintenance of home and yard
  ☐ other: _____

In a new home, would you rather have:
  ☐ a back hall laundry with a smaller kitchen, OR...
  ☐ a larger kitchen with laundry in basement?

continued on next page

In a home with the master bedroom and one other bedroom on the second floor, would you rather have:
☐ a large master bath with separate tub and shower that is shared with another bedroom, OR...
☐ a smaller bathroom for each bedroom?

Would you rather:
☐ have your yard maintenance performed by others and pay an association fee for the service, OR...
☐ provide your own maintenance?

Do you need family room space in a finished basement if it is not on the first floor?
☐ Yes   ☐ No

Do you need a first-floor master bedroom?
☐ Yes   ☐ No

Would the following features be high on your list of "wants or needs" for a new home?
Fireplace:	☐ Yes	☐ No
Security system:	☐ Yes	☐ No
Whirlpool bath:	☐ Yes	☐ No
Cathedral ceilings:	☐ Yes	☐ No
Air conditioning:	☐ Yes	☐ No
2-car garage:	☐ Yes	☐ No

Would you be willing to pay $1,200 more for 2" x 6" exterior walls vs. 2" x 4" exterior walls if it meant your monthly heating bill would be lower by $7.00 per month?
☐ Yes   ☐ No   ☐ Don't know

Would you be willing to pay more for 3/8" plywood construction vs. 7/16" oriented strand board construction?
☐ Yes   ☐ No   ☐ Don't know

How many bedrooms would you use?
☐ 1   ☐ 2   ☐ 3   ☐ 4

In a home with the number of bedrooms you chose in the previous question, how many bathrooms would you expect to have?
☐ 1   ☐ 1&$\frac{1}{2}$ ☐ 2   ☐ 2&$\frac{1}{2}$

Which *two* lifestyle factors would most influence your decision to buy a home?
- ☐ Conforms to my lifestyle
- ☐ Provides security
- ☐ Provides privacy
- ☐ Establishes roots in a community
- ☐ Avoids dealing with landlord
- ☐ Essential to raising a family
- ☐ Better schools
- ☐ Prestige of owning a home

What would you use extra bedrooms for?
- ☐ Guest bedrooms
- ☐ Den/library
- ☐ Hobby/sewing room
- ☐ Computer room/office
- ☐ Children's study/playroom
- ☐ Exercise room
- ☐ Media room
- ☐ Storage

Which of the following most closely describes the kind of home that you would buy, if you were in the market?
- ☐ A house that I would be able to live in regardless of my age or family size.
- ☐ A house that fits my immediate needs.
- ☐ A house that would be designed for expandability to meet future needs.

Which appliance manufacturers produce the highest quality equipment, in your opinion?
☐ _____   ☐ _____
☐ _____   ☐ _____

Which furnace manufacturers produce the highest quality equipment, in your opinion?
☐ _____   ☐ _____
☐ _____   ☐ _____

Which material would be your preference for flooring in a foyer?
- ☐ ceramic tile   ☐ hardwood floors
- ☐ carpet   ☐ vinyl

continued on next page

On a scale of 1 to 5, 5 being highest, how important is having:

Cathedral ceilings:	1	2	3	4	5
Two sinks in master bath:	1	2	3	4	5
A formal dining room:	1	2	3	4	5
A wood deck:	1	2	3	4	5
Rear yard privacy:	1	2	3	4	5
Large foyer closet:	1	2	3	4	5
Mud room/laundry in back hall:	1	2	3	4	5
A ceramic tile shower vs. a fiberglass shower	1	2	3	4	5

For which of the following homesites would you be more likely to pay a premium?
☐ one that is on a cul-de-sac
☐ one that is on a private drive
☐ one that backs up to a pond or lake
☐ one that backs up to trees or woods
☐ none of the above

Do you plan to buy a home within the next five years?
☐ Yes ☐ No

In what month are you most likely to shop for a new home? _____

Your answers will contribute to better housing in your area. Thanks again for your time and effort.

(A slightly shorter version of this survey is used as a mailed survey to consumers.)

This was something they liked. They also liked our house plans. In particular they liked the different ways we configured flow between kitchens, dining, and family rooms.

We also hold focus groups composed only of real estate agents. Generally, agents who have responded to our mailed surveys are invited to participate. If they have returned the survey, we send them a letter thanking them for their input and inviting them to participate in a focus group luncheon. We follow this up with a letter of confirmation specifying the time, date, and location of the meeting.

This letter includes the discussion points for the focus group. Typically ten to twelve agents will participate in one of these groups.

Meeting arrangements for this third type of focus group are much the same as for the other groups. The major difference is that Craig now facilitates these meetings.

We think it is a good idea to use both mailed surveys and focus groups. The surveys provide an overall picture of trends in buyer expectations. The focus groups give us deeper insight into what some of these trends really mean to buyers.

Listening During the Sales Phase. Listening to buyers begins at the initial sales contact. At this point we are paying careful attention to what they have to say about their lifestyles and architectural preferences. Our intent is to determine which of our communities, architectural designs, and price ranges best fit the needs of the prospective buyers. When a successful match is found, the buyer reserves a lot.

Listening During the Design Phase. Our next step is to schedule a meeting with an architectural designer, the buyers, and Craig. From this point the designer begins work on preliminary drawings and the buyers continue to work with Cannon Development to identify specific requirements for their home. Sometimes we discover that what the buyer wants does not match our current specifications. When this occurs, we first look to our contractors and suppliers to see if the specifications can be changed. If so, the change is made. If not, I return to the designer to see if he can suggest changes. Seldom have I had to ask buyers to change their mind.

After entering into a contract we use worksheets like the one shown in Exhibit 3-3 as a way to keep track of the color and material decisions made during this time. The worksheets, once completed and signed, become a part of our contractual documentation.

We permit design and product changes up to 45 days after signing the contract. This is the same time period buyers are allowed to make or adjust color and product selections. We've come to realize that buyers can and will change their minds once they discuss their new purchase with friends and relatives. We also have accepted changes after 45 days if they had little or no significant impact on the other construction schedule. Changes to door knob selections or

Exhibit 3-3: Design and Product Specifications Worksheet

Buyer's Name _____ Home Phone _____ Job Number: _____

Street Address _____ Office Phone _____ Closing _____

Exterior

Roof Shingle Color _____ Shutter Color _____ Gutters _____ Downspouts_____

Check Siding Type _____ Storms Double-Hung: White

Clad Windows: White   Taupe   Bronze   Grey

Casement Screens: White   Taupe   Bronze   Grey

Soffit and Fascia Color (Vinyl): _____ Exterior Stain Color: _____

Brick Selection

Exterior and Chimney(s) _____ Stone Brick Supplier: _____

Fireplace #1 _____ Hearth: Marble Stone Brick Supplier: _____

Fireplace #2 _____ Hearth: Marble Stone Brick Supplier: _____

Other: _____

Kitchen Cabinets Supplier: _____

Location	Cabinet	Hardware	Top	Edge
Kitchen	_____	_____	_____	_____
Laundry	_____	_____	_____	_____
Master Bath	_____	_____	_____	_____
Master Dress	_____	_____	_____	_____
Family Bath #2	_____	_____	_____	_____
Family Bath #3	_____	_____	_____	_____
Family Bath #4	_____	_____	_____	_____
Powder Room	_____	_____	_____	_____
Family Room	_____	_____	_____	_____
Study	_____	_____	_____	_____
Other	_____	_____	_____	_____

Electrical Supplier: _____ **Floor Supplier:** _____

[Brand] Switches Yes ___ No ___

Wood Floors

Type _____ Direction _____

Color: White _____ Ivory _____ Finish: _____ Rooms: _____

Appliances: Brand _____ Range: Gas ___ Electric ___ Dryer: Gas ___ Electric ___

Color _____ Wall Oven _____ Micro/Convection _____ Trim Kit _____

Cooktop _____ Range Hood _____ Downdraft System _____

Burner Cartridges _____ Other Cartridges _____ D/W _____ Trim Kit _____

Refrigerator _____ Door Panels _____ Other _____

Mirrors

Master Bath _____ Dressing _____ Powder Room _____ Family Bath _____

Other: _____

Medicine Cabinets: _____

Shower Doors

Master _____ Bath #2 _____ Bath #3 _____

Other: _____

Wire-coated Shelving: _____ **Wood Rods with Particle Board Shelves:** _____

Ordered: Brick ___ Plumbing ___ Cabinets ___ Ceramic ___ Appliances ___ Wood ___

Field ___ Supervisor ____ K/E ___ M/G ___ Trim ___ Supervisor ____ Cabinets ___

Signed _____

---

sidewalk configurations can be made almost up to installation. And buyers often tell us how much they appreciate being able to make last-minute adjustments to selections.

Listening During the Construction Phase. Even before we began consciously following a TQM philosophy we made Cannon Development production employees available to meet buyers to discuss problems and questions that arose during construction. Such meetings, scheduled directly between the buyers and one of our site supervisors (or our production manager), occur at times convenient to the buyers (during work hours, after work hours, on the weekend, and so on). Indeed, our quality consultant had an interesting experience in one such meeting while we were building his home.

Ray Nichols, our paint contractor, told Doug Allen that he felt Chuck and his wife had picked the wrong color of stain for the wood tub surround in their master bath. Ray felt the color was too pale and made two samples to illustrate his point. One sample was stained with the specified color; the second was stained with a color Ray thought would be more appropriate. Doug arranged an on-site meeting for Ray to make his demonstration to Chuck and his wife. Everyone agreed Ray was right and a change order was initiated.

While informal opportunities to discuss the customers' level of satisfaction are very valuable, scheduled walk-throughs are also important. Our production manager takes the buyers for a walk-through just prior to drywall installation to discuss structural items and mechanical openings. Then just prior to closing, he takes the buyers for a final walk-through of their new home. During this home orientation Doug demonstrates equipment operation, electrical, plumbing, and HVAC features, and reviews Cannon's and the new owner's maintenance responsibilities.

While conducting a home orientation, Doug prepares a punch-list of items that don't match the new owner's expectations for finish. Once the necessary changes are completed to the satisfaction of the new owners, they are asked to sign off on the list. These lists keep getting shorter each year. We believe this is a direct result of our listening more carefully to customers throughout the sales, design, and construction phases and improving how we relate customers' expectations to the trades. In many ways, the builder acts as an information conduit for the home buyer.

Listening During the Service Phase. If problems arise after closing, owners are encouraged to submit a warranty claim form. We ask the owners to hold off using this form for at least one month (unless, of course, something serious happens). Eleven months after closing we contact the buyer, tell them their warranty is about over, and ask them if there are warranty items in need of attention. If there are, we ask them to use the claim form. Owners often prefer to call their problems in. We allow this, and we respond in some way to any issues the owner raises during or after the warranty period. A section of our warranty document also informs owners that Cannon Development is legally committed to correcting warranty problems. Putting this in writing helps sustain owners' comfort level and trust in us.

We still rely heavily on post-closing surveys. The survey form, which we mail to homeowners six months after closing, asks them to tell us if the service work performed and the products used in their homes have met their expectations. We also ask them to rate the performance of Cannon Development sales, construction, and service personnel.

Doug and I now review post-closing surveys as they are received in the office. If we find any ratings or negative comments that indicate we did not meet buyer expectations, one of us calls the homeowner for a further explanation of the problem. Any corrective action that is warranted is then taken.

While taking corrective action, should we learn anything that calls for modification to our specifications, Doug makes necessary changes to the specifications manual and then makes sure the changes are communicated to the appropriate contractors or suppliers. My job is to make sure the general specifications we use in sales presentations are likewise kept updated. (I will explain more below about how we use these specifications.)

At the end of 1994 we added a second survey. One of our homeowners suggested that waiting for six months is satisfactory for discussing whether homeowner expectations for the quality of products and work were met, but if we really wanted buyer input on the overall experience of working with Cannon Development during the buying and building experience, then I, or a company officer, should do the asking. This person explained that if the president of the company calls, the homeowner knows we are serious about customer satisfaction. Now, two months after closing, Barbara calls each homeowner personally in order to "take a pulse" on their overall buying experience with Cannon Development.

Each year in January I ask our office receptionist to summarize all post-closing surveys sent to us during the previous year. The steering committee reviews these summaries to stay abreast of what buyers are telling us about their homes. The committee also looks for negative trends that may warrant special attention. Once again, if needed corrective action involves labor and material specifications, the production manager makes appropriate modifications to the manual; and I make sure the required modifications are given to future buyers. Chapter Five will include more information on how we now handle post-closing surveys, including our sixty-day telephone interviews.

Listening to Buyers After the Warranty Period. As mentioned earlier, we continue to listen to buyers after the warranty period on their home. Sometimes we find the problems were caused by the owners. For example, a bathtub that begins to clog after five years is usually the result of the owner's failure to clean out the drain periodically. On the other hand, some problems are due to workmanship or material defects that took some time to show up.

One owner called recently complaining about unusual air infiltration around several baseboards during some extremely windy and cold weather we were experiencing. She had been in her home for over nine years and had not experienced this trouble before. What Doug found was a gap between the carpet and the baseboard in the affected areas. Discussing the situation with our insulation installer, Doug learned of a new insulation product that could correct the problem. The distributor recommended Doug try the product and provided him with the necessary materials at no charge.

The solution appears to be working; the customer experienced no problems during our last bout of windy weather. While we were not obligated to perform this work (and the product was not available when we built the home), common sense seemed to dictate that we should at least try to make things better. However, other problems are not solved this easily (or cheaply). Correcting odors emanating from fireplaces due to wind conditions or replacing windows that lost their seal have taken more time and money. In some cases, we cover the complete cost. In others, we share the cost with the owner.

In general our approach after warranties expire is to listen to the complaint and investigate the situation. When the problems appear to be caused by poor workmanship or the use of a defective product, we correct the problem and try to get the expense shared by the distributor or installer. When the problems are caused by the owner, we recommend a contractor who can take appropriate corrective action.

Educating Buyers. While today's buyers are more aware of what to expect from a new home, we enhance their awareness by providing a good deal of education to them throughout the building process. Taking time to do this is a significant aspect of TQM. And while we have always provided a good deal of buyer education, it has most certainly taken on new importance in the past few years. For example, the ad hoc team that investigated our problems with cracked concrete strongly recommended that we educate buyers on the causes and the actions taken to solve this problem. The steering committee accepted this recommendation without hesitation because they knew Cannon Development already had a mechanism in place to handle it. But buyer education is important to TQM in other ways.

By providing buyer education early, we help build trust between the builder and the buyer. People are kept informed from the beginning on why we build homes the way we do and what to expect as we move through construction and service phases. Buyers come to expect we will be forthright when they ask questions later about things they see happening to their homes.

While we do not want buyers to lower their expectations for product performance, we do want to make sure the performance ability of products under consideration matches their expectations. Education is how we accomplish this. For example, I tell buyers that wood floors will shrink in the heating season even with humidification. Explaining how to minimize this shrinking helps buyers make a more informed decision should they decide to use wood. Similarly, if the buyer is debating between insulated and high performance glass, I point out that condensation will still form on high-performance glass when there are extreme shifts in outside temperature and humidity. Knowing such things up front reduces the likelihood that buyers will feel the need to call for explanations once they move into the new home. It also reduces the number of service callbacks.

Product education is too important to leave to chance or whim. After a contract is signed, someone needs to assist buyers in selecting colors, finishes, and materials as well as help them understand that some products perform better than others. At Cannon Development, that "someone" is my wife, Barb.

The documentation Barb finds helpful for educating buyers on product performance is extensive. It includes general specifications for the home being built, warranty specifications and procedures, and product information (including how the products should be maintained). She also keeps wood floor, tile, brick, and paint samples on hand and has established liaisons with our cabinet, electrical, and plumbing vendors. These liaisons are very important to what Barb is trying to accomplish.

Barb identifies one person at each vendor who is willing to serve as an informal representative of Cannon Development on their own showroom floor. She makes sure each liaison is fully versed with how we design our homes and the procedures we use. This makes life for our buyers much easier when it comes time to pick kitchen cabinets, choose light fixtures, or decide on the type of plumbing fixtures for their new home.

Barb spends several hours with buyers on product selections. She will hold meetings with individual buyers at our offices, will sometimes go with them to visit a vendor, and often talks with buyers over the phone during normal work hours as well as in the evenings or on weekends. She finds that using a worksheet (like Exhibit 3-3) helps keep decisions from falling through the cracks. Barb also takes time to explain to the buyers what they should expect to see happen during the construction and service phases of building their home.

The feedback I receive from real estate professionals, former clients, and referrals is that Barb is very good at buyer education and does a lot to help make the buying experience pleasant. As one buyer wrote to us, "We appreciate the extra steps you personally took to work with us as first-time buyers. No other builder was willing and or able to offer us the flexibility that you did." Such feedback leads me to believe Barb's educational program is also an excellent marketing tool for Cannon Development. Satisfied buyers continue to be our best advertisement!

Listening to our buyers over the years has allowed Cannon Development to make many improvements. But we don't always meet (let álone exceed) buyer expectations. Prior to our TQM days I assumed it was my responsibility to identify and fix anything that did not satisfy a buyer. Now I realize the people who work with me to build a home often are in a much better position to tell me when something needs improvement and how the changes should be made. Employees, contractors, suppliers, distributors, and manufacturers have much more knowledge and expertise than I on how to take appropriate action to solve specific problems. Listening to internal customers is critical to process improvement.

Discerning Internal Customers' Expectations

Just as we do with buyers, we listen to employees, contractors, and suppliers in two ways. One way is to solicit their thoughts on where we need to make process improvements at training sessions, through surveys, and at meetings we hold once a year. The second way is for Cannon employees to listen while they are out in the field or when internal customers drop by the office. Regardless of when or where we "hear" of a problem, our approach to converting these inputs into process improvement is the same. Input that indicates problems are complex or involve more than one internal customer will be forwarded to the steering committee for deliberation

---

Tip

When to Refer a Problem to the Steering Committee

In the early days of our journey employees, contractors, or suppliers would occasionally ask for guidance on whether a problem should be handled by themselves or referred to the steering committee for resolution. Until people felt comfortable with making such a decision, I told them, "When in doubt, refer a problem to the steering committee." The committee then decided who should handle the situation. In no instance, however, did the committee ever take on an issue they believed would be better handled by the person who brought the problem to them.

and follow-up. Problems that involve only one or two internal customers are typically handled by Doug or myself.

Listening by Asking. Asking employees, contractors, or suppliers to identify process problems seems simple enough. But at first people can be hesitant to raise specific issues. For example, when I asked for this input during the initial TQM training I received no responses other than confirmation of the list the steering committee had already developed. My readout of this situation was that people were waiting to see what would happen before raising their own issues. Silence does not necessarily mean problems don't exist. I suspected they were uncomfortable about talking about their problems in front of people they might perceive to be indifferent or hostile.

We overcame the "wait and see" attitude by widely communicating each success we experienced with process improvement and by publicly praising the people responsible. During the first years of our TQM journey we overcame contractors' and suppliers' fears of being seen as "tattlers" or "crybabies" by using a written survey to collect their input. This mechanism allowed people to get things off their chests without having to do so publicly in front of a group. It gave Cannon Development a much clearer picture of concerns held by our internal customers.

While the survey worked for us, I felt a more efficient way to collect input from large numbers of internal customers quickly would be to ask for it during our annual supplier and contractor meeting. For years, Cannon Development has held such meetings with employees, contractors, and suppliers to discuss mutual business concerns. But until we had built a comfort level with TQM, I did not think talking about process problems during this meeting would go anywhere. By the early 1990s, however, our journey was well under way. I decided it was time to use the annual meeting to ask participants to identify process-related problems.

As chair of these meetings, I strive to create an atmosphere in which people are willing to discuss what's on their minds. I accomplish this by listening, by not criticizing or demeaning suggestions, and by following up in a timely way.

Asking direct questions about process problems in surveys and at large group meetings is an efficient way to obtain people's input. But to me, it's something of an artificial process. I see various internal customers on a daily basis and it seems more natural to talk with people about process problems now rather than waiting for a meeting or the next survey. This is easy enough to do with employees I see every day in the office; it is more difficult to do with people out in the field. But after more than six years' experience with TQM, the vast majority of process problems surface during daily interactions between Cannon Development and our internal customers. Many management consultants call this "management by walking around." I call it "TQMing."

Listening While TQMing. One of our production employees visits each construction site daily to check on work progress. We look for a variety of things: Are the materials present to complete all the work scheduled for the day? Have damaged or unacceptable materials been set aside for return? Do site conditions warrant accelerating or delaying materials delivery for the next trade? We listen to what contractors and suppliers have to say about how well their expectations are being met and demonstrate by our presence (and later by our responsiveness) that we care about what is important to them.

A word of caution: Mere presence on a construction site does not guarantee we will hear what we need to hear. Nor does it demonstrate that we recognize contractors' or suppliers' concerns. Contractors in particular can be hesitant to bring up problems. We have had better success by asking them direct questions about how they found work conditions. This type of questioning clearly invites them to tell us about their problems. We take the same approach when a contractor or supplier comes to our office. We let them know, by asking questions, that we are interested in their point of view and want to know when their expectations are not being met.

Regardless of where or how we learn about problems while TQMing, our approach to converting this information into process improvement is the same. First, we discuss the issue with everyone involved to make sure we all understand the problem. Then we talk about solutions—who they will affect, whether there are alterna-

> **Tip**
>
> **Updating the Specifications Manual**
> Our production manager keeps our specifications manual updated. Rather than make changes each time a modification is called for, he puts a yellow self-sticking note on each page that requires editing. Then, every six months or so, he incorporates all the changes at once. He finds this an efficient way to keep the manual current.

tives, and who should make the change. As production manager, Doug makes sure any agreed-upon solution that effects material or labor standards is noted as an update to the specifications manual. If the solution involves a change to a contractor's or supplier's business practices, that contractor or supplier is responsible for communicating the change to their employees. Here's an example of how TQMing works for us.

During one of my site visits an electrical contractor alerted me to a problem he was having with switch boxes. At some point after he installed the boxes, they usually became filled with drywall compound. On finish, the electrician would knock out the dried compound, which sometimes caused a rip in the drywall below the box. The drywall contractor then had to come back to make repairs. In addition, the compound would fall onto the floor and be left for someone else to sweep up.

Doug and I brought the electrician and drywall contractors together to investigate the situation. We found the problem typically occurred when a tape joint occurred around the box opening. We decided that in the future switch boxes would be placed at least six inches below tape joints, a minor change to our existing specifications. The drywall contractor assumed responsibility to monitor the solution and agreed in the future to clean out any electric box openings he found full of drywall compound. This has solved the problem. Now the electrician spends less time installing switches and the drywall contractor doesn't have to make repairs so often. And we have less mess to clean up.

Some problems we learn about through TQMing involve several contractors or suppliers. Or, a problem can be very complex or lack a readily apparent solution. When either situation arises, we refer the problem to the steering committee for action. The committee then sets up an ad hoc team to investigate the issue and recommend how it should be resolved.

A few years back several contractors complained about our policy on buyers visiting the construction site. The contractors noted that the buyers' presence during work hours was disruptive. An ad hoc team investigated this issue and found that, too often, buyers engaged the crews in long discussions about the house. The team recommended that buyers should still be free to visit the sites. To mitigate the distraction, however, they recommended that the trades politely refer all questions outside their specific knowledge to Cannon Development. The steering committee felt this was a reasonable solution and it became our formal policy on buyer site visitation.

Occasionally the steering committee will handle a matter by itself rather than referring the problem to an ad hoc team. Because the roles of the steering committee and the ad hoc groups are distinct, this is done only when the committee feels confident the solution is obvious. For example, several of our contractors and suppliers used to complain about difficulty in reaching our production personnel when they were out of the office or field trailer. After discussing the issue, the committee recommended that supervisors carry mobile phones and digital pagers. I carried out the recommendation with one slight modification. To reduce costs, we use the pagers only to alert supervisors to incoming calls.

These examples may seem simple, but they are typical of how the needs of internal customers can be translated into process improvements. When we put together the numerous small improvements we have made over the years, we realize our problem-solving process has become significantly more streamlined and cost effective than it was in the early days of TQM.

Contractors and suppliers who were reluctant to bring up problems in the early days have largely changed their attitude. They now volunteer information about job conditions. But active listening and prompt responses continue to be important.

Over time, as everyone has become comfortable with our way of handling problems, reliance on ad hoc groups has given way to a more natural, informal relationship with our contractors and suppliers. We still use ad hoc groups, however, when the situation warrants.

The Value of Information

Information leads people to the resources they need to make process improvements. Until we got into TQM we stored much of our information in our heads. It was difficult to share information without distortions creeping in. Now, with a specifications manual, written general specifications, architectural drawings for each home, and forms to collect input from buyers, we can ensure that everyone is using the same information to make decisions.

Documentation also helps us monitor how well we are meeting our goals. However, I believe data used for this purpose need not be highly statistical. Simple charts, graphs, and percentages can reveal a lot about a process that needs to be fixed—or indicate that things are working as planned.

Tip

Compiling Meaningful Reports

With all the data lying around our offices, I am tempted to compile all kinds of reports. I fight this temptation by reminding myself of a simple rule of thumb: *Reports should tell you something useful.* I avoid creating reports just for the sake of having them around (or just because some data are available). In a similar vein, there is a shelf-life to the reports I do keep. I don't keep them around forever—something we fully appreciated recently when we moved our office.

Much of the information needed for process improvement probably already exists in your company. It does for Cannon Development. We have not found it necessary to create *new data*. The key has been to *organize* existing information. Also, we believe a report should tell you something useful. If it doesn't, the report's not needed. The following sections illustrate how we manage information at Cannon Development.

Tracking Time

One measure we use for process improvement is the amount of time it takes to complete the final walk-through service. A reduction in this time indicates that much, if not all, construction went as planned. But simply tracking service completion time is not enough. To take immediate action on problems that arise during construction we also need to track time throughout the construction process. Here are some examples of how we track time and use this information to make process improvements.

Weekly Schedules. Assume that on June 7 I receive a signed contract and all appropriate attachments to build a home for the Smiths. Within 24 hours I will authorize Doug to start construction. At this point Doug does two things. First, he adds this start to the schedule of activities for the week he plans to begin construction on the new home. Let's assume the lot stake-out is to occur the following week and that other preliminary work will be completed by the end of the month. Doug will thus have three weekly schedules with dates showing when various types of work will begin and end on the Smiths' home.

Doug's second task is to complete a series of checklists we have designed to make sure everything is set as each phase of construction is initiated. (These lists resulted from a meeting with contractors in which they raised concerns about not being ready when they arrived to start work.) As you scan the sample checklists shown in Exhibit 3-4, note that many items refer to work which must be competed before another phase begins.

Doug (or one of our other field supervisors) calls the appropriate supplier or contractor to confirm job progress and arrange the detailed job schedule. Doug makes follow-up calls, usually about four days in advance, to confirm all is set; then he meets with the contractor or supplier the morning work begins to discuss the job situation. The checklists serve as a reminder of all the things Doug must check on before work begins in each phase.

Back to the Smiths' home. Work on their house starts in June. Doug creates weekly schedules for all work to be initiated during this month. Throughout the building cycle, he will continue to create weekly schedules that indicate what trades will be on what jobs

Exhibit 3-4: Supervisors' Checklists

504 Surveying

☐ Job date _____
☐ Change orders	Yes	No
☐ House plans to surveyor	Yes	No
☐ Lot cleared for stakeout	Yes	No

517 Rough Grading

☐ Start date _____
☐ Completion date _____
☐ Change orders	Yes	No
☐ Foundation tarred	Yes	No
☐ Foundation inspected	Yes	No
☐ Foundation braced	Yes	No
☐ Block pickup scheduled	Yes	No
☐ Stone ordered	Yes	No
☐ Sand ordered	Yes	No
☐ Fill required	Yes	No
☐ Electric cable ordered	Yes	No
☐ Phone cable on site	Yes	No
☐ Television cable on site	Yes	No
☐ Grading plan reviewed	Yes	No
☐ Drive location reviewed	Yes	No
☐ Electric location reviewed	Yes	No

552 Rough Electric

☐ Start date _____
☐ Completion date _____
☐ Change orders	Yes	No
☐ Kitchen plans done	Yes	No
☐ Electric meter installed	Yes	No
☐ Attic fan installed	Yes	No
☐ J-blocks installed	Yes	No
☐ Security system	Yes	No
☐ Basement floor poured	Yes	No
☐ Basement framed	Yes	No
☐ Garage floor poured	Yes	No
☐ Trim sizes	Yes	No

and on what days. This helps us avoid scheduling conflicts. We use the same schedule to assure work is completed on each job before the next crew arrives.

Long-Range Schedules. During the first week of each month Doug adds all starts from the previous month to a long-range schedule. He also indicates when he expects critical phases of work on the Smith house to be completed over the next four months. The schedule includes space for Doug to note the time it takes to complete work on other homes during this same four-month period.

As the months go by, Doug notes on the long-range schedule whether or not we met the expected schedule of activities from the prior month. He identifies the jobs that have dropped behind schedule and those that are ahead of schedule. Doug and I track this movement so that we can keep buyers informed of any changes to the expected delivery dates stated. Buyers appreciate having up-to-date information, especially as they prepare for closing. Our record for meeting the delivery date stated at contract signing has dropped from a little over three days (before TQM) to less than a half day.

Construction Time Variance. Information on the long-range schedule alerts Doug and me to scheduling conflicts and helps us track the progress of each home from month to month. This information doesn't tell us how much variance we have in estimated versus actual time to complete each phase. We use a bar chart to depict this variance. By comparing phases in bar chart form we can easily identify which trades continually fail to meet their own time estimates for work.

Tracking Construction Costs

Another of our quality goals is to reduce the variance between estimated and actual costs of construction. We use annually quoted prices to estimate the cost of labor and material for many phases of construction. Estimates for materials also may be based on recent historical data. We use a spreadsheet to track the variances between the estimates and actual costs for each house.

Each invoice for labor and material used is approved by the field supervisor. Any invoice over 3 percent of estimate requires the production manager's review. Invoices with at least a 10 percent variance from estimate are brought to my attention. Explanations for overages are written on the invoices. Later, when we're looking for patterns in cost averages, we don't have to rely on memory.

At the end of each year Doug and I review all invoices to identify variance trends and to develop a big-picture view of how well we are containing costs. But the best time to take corrective action is when each invoice is compared to the estimate as described above. I'll have more to say about this process in Chapter Five.

Tracking Business Functions

Any member of the steering committee, an ad hoc team, or our office staff can access the information described above. As we stated earlier, I have also promised any ad hoc team access to financial information to assist them in solving process or specification problems. And, as will be explained in the next chapter, Cannon Development employees also use such information to keep abreast of our bonus plan.

Financial Information. Cannon Development's overall financial picture is summarized in a standard profit and loss statement on a computer-generated spreadsheet. This report tracks all direct and indirect costs incurred monthly. It also shows a ratio of actual expenses to budgeted amounts as well as year-to-date expense figures. While there is nothing special or unusual about how we keep this report, it is important to understand that it is a report *all* Cannon employees help put together. It is also a report to which everyone in the office has access.

Use of the Computer. Most of the reports discussed in this chapter are computer generated. And we are currently investigating ways to produce all our reports, especially those we use for scheduling and time variance, on the computer. From prior experience in making similar conversions, we know it is important to first recognize what kinds of information, summaries, and displays we need. The computer cannot make those decisions for us. But we haven't given up yet.

Collecting Information While TQMing

As explained earlier, issues that involve only one or two trades generally do not require multiple meetings or submission of final reports. Our common goal is to reduce time and save money, not sustain special committees. Here's an example to illustrate what I mean.

Our painter numbers each leaf of the door hinges before pulling the pins and removing the doors for painting. Afterwards, the trim carpenter re-installs the hardware and mounts the doors. Our cleaning people then remove the numbers the painter makes. Since our cleaners work on a fixed price, the quicker they can get their work completed to specifications the quicker they can get on to another job. (The quicker they are, the more money they make.)

Recently, while Doug was walking through one of our houses with the painting contractor, one of the cleaners remarked how much time they were spending cleaning black marks off door hinges. The cleaner then asked if the painter would mind switching to a water-based marker. The painter said, "No problem." And that's what he now uses.

So what does this all have to do with information and TQMing? The point is simple: The cleaner's problem was solved in about as much time as it took you to read about it. Our approach to process improvement while "TQMing" is to walk around, listen, discuss, and then act. Checksheets, bar charts, forms, and committee reports are seldom, if ever, needed. They would only get in the way. Bottom line? "TQMing" need not require cumbersome formal reports.

TQM Is An Attitude

TQM affects everything we do as a business. It is not something we practice every now and then. Listening to customers and converting what we learn into process improvements has become habit for everyone at Cannon Development. While we stumble at times, stumbling does not mean we're off course. Each challenge we overcome motivates us to try even harder. To us, this is the aspect of

TQM that makes it a never-ending journey. We are never totally satisfied. There are always better ways to build homes and satisfy customers.

We of course did not get to this point in our journey overnight. Before we even started the trip, we all had to do a lot of planning (and learning). We deliberately took our time during the early days of the trip in order to make sure we knew what we were getting into. But while thorough planning and starting slow created a great deal of momentum, neither were sufficient to explain what kept us on this road for more than six years. The next chapter explains how we have been able to sustain our journey.

Chapter Four

Sustaining the Journey

Listening to customers and converting what we hear into process improvements are important components of TQM. But I have found that listening and adapting to improve customer satisfaction are not by themselves sufficient to sustain our journey. This chapter explains additional steps we have had to take.

People are the most important component of any business strategy for a home builder. People buy houses built by people. While our business (and our TQM efforts) would not exist without buyers, our internal customers are the "engines" that drive TQM and our business. We must therefore take steps to keep our engines well tuned.

As illustrated in previous chapters, better customer satisfaction and cost containment stem from listening to customers and responding to what you learn. The ability to listen is also critical to sustaining a TQM effort. Learning to be better communicators kept us on course.

With TQM life has not always been smooth. We encountered some significant bumps in the road during our journey. And as you will see below, some of these bumps could have brought our trip to a halt had we not done something about them.

Keeping Our Engines Well Tuned

As owner of the company, I believe is my responsibility to make sure all employees, contractors, and suppliers receive the resources and support they need from Cannon Development to help us achieve our goals and promote a change in our culture. My goal is for the TQM attitude I've talked about in this book to become everyone's way of thinking and acting on the job. This requires a significant investment in our people. But it is an investment well worth making.

Training and Education

There is no one right answer—no "cookbook"—for promoting learning. But what we do works for us. Our focus is on *learning* rather than on training and education. Learning is a change in behavior (listening better; accepting responsibility to act on what we hear; and using a more systematic, objective approach to problem solving). Training and education are just two different ways adults learn. We also learn through experience. We try new things (such as ad hoc teams or consumer focus groups). If they work, we keep using them. If they don't work, we try something else. Training and education can help us learn *about* new techniques. But learning *how to use* the techniques really happens when we try them out. Likewise, training and education do not teach new attitudes as successfully as experiencing the satisfaction that comes from meeting or exceeding objectives.

Keeping the Steering Committee and Ad Hoc Teams Up-to-Date. Once every year or so, Chuck holds a mini-workshop for the steering committee. These sessions typically are 2 hours in length and cover such topics as how to collect and analyze data, how to measure the cost of nonconformance to specifications and standards, and how to benchmark. This training serves as both a refresher for the committee and a way to address special topics. We have not had to retrain the entire committee on the basic tenets of TQM.

In addition to holding the workshops, Chuck meets once or twice a year with individual (or a few) committee and ad hoc team members to discuss specific issues related to problem solving and running meetings. These informal sessions usually are held at the request of the committee or team member. We have not found it necessary to provide more formal training for ad hoc teams.

The cost of providing this training is modest. Typically, less than $250 will cover instructional fees, materials, and refreshments for one of the steering committee workshops. It could easily cost three or four times this amount to send the entire committee to a local seminar on the same topic (and where the material may not be geared specifically to home building, let alone to our specific situation). No direct expenses are associated with the informal get-togethers for individual committee and team members.

Orientation for New Committee Members and Internal Customers. Our steering committee has remained pretty much intact over the years. When someone new joins the committee (we have only had two new members to date), the committee chair (Doug Allen) gives the new member a copy of the materials we used during our initial training and previous mini-workshops as well as copies of committee minutes and past ad hoc team reports. He asks the new member to read over these materials and then spends an hour or so going over them. This is about all it takes to get a new member up to speed.

Doug finds the specifications manual an excellent educational tool for explaining construction standards to new contractors and suppliers. He also finds sharing training materials used in the early days as well as various ad hoc team reports helps bring the new vendor up to speed on how we approach problem solving and process improvement. But nothing illustrates TQM better to a new person than having the people who have participated in past process improvement activities explain what they accomplished from their point of view. Success stories—told by the people who did the actual work—help greatly in orienting new people. Such stories are both illustrative and nonthreatening. The easiest way to accomplish this is for the new member to sit in on a steering committee meeting where an ad hoc team is making a report. When this is impractical Doug will ask employees, contractors, or suppliers involved with recent projects to make a presentation at our annual meeting.

Continuing Education. I try to stay abreast of trends in TQM by talking regularly with Chuck, reading professional journals, and attending conferences. Any books or journals I pick up are passed around the office and to the steering committee. I also pay for employees to attend local residential construction seminars or other continuing education programs. Several of our employees have attended OSHA seminars and local quality conferences. Our office accountant has taken cost accounting and computer application courses at our local community college. All that I ask is that the educational program be somehow relevant to the person's job.

Hiring Practices

Until TQM becomes more widespread in the industry it's unlikely that new hires will be able to catch on immediately. We therefore provide all new hires a thorough orientation to what TQM means at Cannon Development. Even so, people already in our organization have to be patient while the TQM way of thinking sinks in. We have found that a person's predisposition toward TQM principles can affect how quickly they adapt to our way of doing business.

We look for three characteristics in any applicant for an open position at Cannon Development. First, the person should have work experience and feel comfortable in a team environment. This person should also be willing to serve on the steering committee or an ad hoc team. During the interview we listen to the way applicants talk about current or former jobs. If the word "we" is used a lot, they probably see themselves as team players.

Second, we look for a person who has used data and information to make decisions. Asking applicants to explain how they made a recent decision—either a personal one or a professional one—is a good way to get at this. Someone who can articulate how conclusions are reached, and on what basis, is most likely a quantitative decisionmaker.

> **Tip**
>
> **Training and Education**
>
> I find that people learn more when training and education is provided in small doses. And training works best if it is provided only to the people who need it and only when they can apply what they learned. Training and education efforts "stick" when they are provided "just in time."

Finally, we look for people who prefer to solve problems rather than point fingers. To test for this, we ask applicants to explain what they would do first to solve a problem. If the first words out of their mouths are in effect to find out who is to blame, they are finger pointers.

We typically replace one to three contractors or suppliers each year. With the team in mind, we interview potential new contractors in pretty much the same way described above.

Recognizing Internal Customer Contributions

Nothing seems to promote pride and a positive attitude about work more than public recognition of our internal customers' contributions. We try to take every opportunity to let everyone in the office and the field know when something significant has been accomplished. The flip side of pointing fingers of blame is pointing to accomplishments or to those individuals who have exceeded expectations. Public acknowledgment helps fuel people's investment in quality work.

Recognizing Companywide Accomplishments. When Cannon Development won its National Housing Quality Award in 1993 Barb and I wanted to recognize the efforts of our employees. It was their hard work and contributions that had made this possible. So we decided to take all our employees and their spouses with us to the award ceremony in Washington, DC. We not only wanted to thank them, but to thank them in the presence of others at the meeting.

We also wanted to recognize the efforts of our other internal customers in helping us win the NHQ award. We thought a good way to do this would be to give contractors and everyone (regardless of their level of involvement with us) a special tee shirt at our annual picnic. The shirt has "I won the NHQ award" imprinted on the front. To this day, I see contractors wearing that shirt. It is a source of pride to them, and great advertising for us!

We use gifts such as tee shirts to recognize any companywide accomplishment. And whenever we decide to give out such gifts, we make sure everyone gets one. We want to be sure we have not overlooked someone.

Recognizing Individual Contributions. We avoid going overboard with special plaques and bulletin board notices as a way to recognize an individual's contribution. However, we do have a plaque in our office reception area that recognizes past and current winners of our "Best New Contractor," "Best Contractor," and "Best Supplier" of the year awards. In addition, I have a plaque on the wall in my office that recognizes recipients of our "President's Award." This award identifies a contractor who has continually demonstrated an ability to meet production service schedules while at the same time continuing to raise personal standards of quality work. Feedback I receive from everyone in our extended organization indicates all these ways of saying "thanks" to individuals are appreciated and seen as sincere.

Cannon Development Bonus Plan. Our employees recently proposed, and I accepted, a bonus plan tied to our quality objectives and annual net profits (before taxes). Since everyone in the office contributes to our overall performance, everyone receives a bonus (except Barb and I—our bonus as owners is increased retained earnings). The amount paid for meeting quality objectives is paid quarterly (assuming objectives were met).

We also give an annual bonus. The dollar amount of this bonus is based on a percentage of total profits over an agreed-to base amount of profit. Our employees understand that a certain amount of profit must be added to retained earnings if we are to continue to grow.

Our two bonus plans have not been in effect long enough for us to tell just how well they are working out. But I can tell you our employees seem pleased that I am at least willing to give this form of recognition a try.

Sustaining Employee Morale. Providing resources and support for employees to do their work goes a long way in promoting morale around the office. Being responsive to their other needs also is important. We have flexible work hours, allow job sharing, and pay for a comprehensive health plan. We do not do this because I am a benevolent business owner. We provide these benefits because they address employees' needs, promote good morale, reduce turnover, and help people focus on their work.

Listening to employees also promotes high morale. The bonus plans noted above arose from employees helping to create the company's annual financial and quality goals. Since they have a say in setting and achieving these goals, they felt they should share in the success.

Employees also have lots of good ideas on how to improve life around the office. For example, three of our staff took responsibility for determining how space should be utilized in our new offices. Similarly, the two employees who answer our phones asked to be kept up-to-date on sales and general design information for our new developments. They are often able to answer inquiries without having to refer callers to another person. This allows callers to determine immediately if they wish to pursue building with us. When you stop to think about it, making contributions to life around the office is just as much a part of TQM as making contributions to life on the construction site.

Staying on Course

I believe our ability to incorporate TQM into the way we think about building homes is a direct result of our ability to listen. Communicating with buyers has always been one of Cannon Development's strengths. The steps we take to listen and respond to buyers throughout the building process are not new ways of doing business for us. What is new about this is the amount of information we now have in written form. Hearsay and memory loss are no longer issues for us. The real challenge for us has been to improve communication with our internal customers.

Communicating with Internal Customers is a Two-Way Street

Over the years, I have often found that builders tend to communicate with employees, contractors, and suppliers by "telling" rather than "asking." There appears to be a widely held misconception among builders that we know more than the other people with whom we work on a daily basis. While it was never my intent to come across this way, it apparently was the way I used to be perceived by our internal customers.

Building Trust with Internal Customers. By learning to do more asking than telling, I have found internal customers are indeed very interested in doing good work and have a great deal of pride in what they accomplish.

Our internal customers now realize I am not out to "get" them. They know I am only seeking their advice and help on how to improve the home building process. We had a real breakthrough when internal customers discovered I was asking them questions they never thought I would ask (or care to ask). Too often, my earlier lack of questions had been interpreted as my not caring about how they found site conditions or about how they went about their work.

Internal customers are now more willing to share their thoughts about improvements Cannon Development can make. They are also more willing to talk about how their own companies can improve. Open and frank discussions about how to improve "things" rather than "you or me" builds the trust that is necessary for true two-way communication to occur. But trust is also built in another way.

Once I started asking questions, the doors to many subjects were opened. Cannon managers were now able to talk with other business owners and managers about a myriad of issues such as OSHA, equipment costs and maintenance, workers' compensation insurance rates, and employee turnover. Through these many discussions we have come to realize we share with our contractors many of the same issues and problems. This understanding may not solve specific problems or change work habits, but it does create a bond among internal customers and Cannon.

Maintaining Trust. As Cannon Development has improved our lines of communication with internal customers we have found it necessary to use a great deal of discretion and common sense to protect our new-found trust. When one of us points out a failure to meet work or scheduling standards we are careful to state our criticism in terms of quantifiable evidence (such as a variance to the standard agreed to up front) rather than in terms of their attitude. By avoiding questions like, "Why did you do that?" or "Don't you know better?" we keep trust from breaking down. I try to focus on the issue, not on the individual. Otherwise, our internal customers will not take risks, try new techniques, or remain cooperative.

As trust continues to grow, everyone realizes they have something of value to contribute. This goes a long way toward reinforcing a positive team-oriented attitude. Trust also reduces suspicion others may have about my intentions for TQM. Such outcomes are too important to leave to chance; therefore, I place a great deal of importance on communication as a way to build and then maintain trust.

Eliminating Barriers to Implementing TQM. I try to communicate to internal customers that we have two common goals. We all want to please the buyers of the homes we build. And we all want to make money. Otherwise we won't last in the industry. At the same time, I try to communicate my desire to break down the barriers that keep us from doing this.

Suspicion and mistrust—the two most common barriers—are rooted in the result of life experiences both internal and external customers have had, or continue to have, with builders. Contractors may have been paid poorly, late, or not at all by some builders. Crews sometimes arrive at jobsites where it is impossible to work efficiently. Suppliers attempting to deliver materials find the builder has provided no access to the jobsite. Such experiences create suspicion and mistrust in builders' internal customers. And truly negative attitudes are generally the result of many years of negative experiences with builders. I thus do not expect suspicion and mistrust to disappear overnight. But listening to and cultivating positive relations with internal customers yields great rewards in the long run.

Listen, Listen, Listen

We only perceive a person's needs by truly listening to what that person has to say. If that person is upset about something, however, what we first hear tends to be frustration and anger. We may not actually learn what the problem is until the person has had a chance to blow off some steam. And waiting to listen sometimes requires considerable patience.

The person we are listening to may also be very disorganized in what he or she is trying to say, especially if the person is upset. As I listen to someone talk, my tendency is to form judgments about what is being said and jump quickly to a conclusion. When I do this, I'm not really listening. This can get me into trouble, as my judgments and conclusions may not be correct. I have to work on remaining patient and letting the person talk things through without interruption.

This is, of course, much easier said than done. But I *can* do it when I remain patient and attentive. If I allow myself to become defensive or personalize the situation, listening—and communication—can break down. By contrast, if I am patient in the beginning, I find we soon are able to get on with clearly identifying the root problem and finding a solution.

Communication with Office Staff. Communication with office staff comes a lot easier when it occurs in an informal, relaxed atmosphere. However, it can be hard to create such an atmosphere during normal work hours. One of the things we do to get around this is celebrate birthdays with a lunch in the office. All employees are included. During these informal gatherings I usually give a brief update on how things are going with current jobs, the company as a whole, and where I see us headed. I also ask each employee to give his or her own update and forecast for how things are going with them. Everyone tells me they appreciate the informality of these discussions and that they find the information helpful.

Communication on Jobsites. I make it a habit to visit each construction site periodically just to say hello and engage those I find on the job in discussions on topics of interest to them. Maybe we will talk about the new truck, how kids are doing in school, or if the Buffalo Bills will ever win a Super Bowl. I am not there to raise problems or take corrective action. The tradespeople should see me out on the job without worrying about why I am there. I am there to hear about issues of interest to them. There is a downside to this technique, however, that should at least be mentioned.

At times I come away from my informal site visits with the feeling those with whom I spoke honestly feel I will be able to solve all their problems. Of course this is not the case. There I many things I can't do. However, I do come away with things to think about and will at least attempt to find answers. But even if I can't solve the problem, I have learned more about life on the job and maintained an open channel of communication.

Written Communications. We use a one-page flyer to keep all our suppliers and contractors abreast of current news at Cannon Development. This also gives me an opportunity to update everyone on ad hoc team and steering committee activities. Also included periodically are summaries of joint meetings we hold with employees, contractors, and suppliers. The flyer is usually distributed in employee and vendor pay envelopes and we mail copies to those contractors and suppliers who may not be working with us during the pay period.

We also use written communication extensively with external customers. In addition to the documentation given to buyers throughout the building process, we recently started mailing a quarterly one-page newsletter to all our homeowners. This letter keeps owners informed of builder-related activities going on around them. We include the names of new residents, updates on product information, yard and maintenance tips, and other items that may be of interest to buyers. Copies of current newsletters are also given to sales prospects.

The informal and formal communication techniques discussed here work well for us, but not because we "know how to communicate." I believe they work because people recognize our sincere desire to help everyone achieve their goals and stay in business. Cannon Development staff "walk the talk." We also are persistent in our efforts to break down barriers created by suspicion and mistrust. Keeping on course is not so much a function of *how* we communicate as of *what* we communicate and our consistency in making the effort.

Bumps in the Road

The TQM journey is not always smooth. Employee, contractor, and supplier enthusiasm was high in the early days but began to wane over time. Changes in personnel within this group meant at any given time different people were at various points on our TQM itinerary. Empowering people sounds nice, but not everyone wants to be empowered. And, most recently, the market for new homes in our area went dreadfully sour. We successfully navigated around these bumps in the road by first recognizing them and then making appropriate mid-course adjustments. Here is what we have done.

Maintaining Enthusiasm

Early successes by ad hoc teams sustained a high level of enthusiasm in the early days. Other tangible signs of improvement, such as written material and labor specifications, also helped. As time went on, however, I believe the little changes that began to occur in everyday life on the job were what kept us all going. My task became to ensure that these small successes were communicated widely. And, as discussed previously, it is always better to let the other people involved do the telling.

People are motivated by different things. People who are motivated by fear will work hard to avoid *what* they fear: losing their livelihood. But fear seldom motivates people to achieve their fullest potential. Taking responsibility to find better ways of doing things can be risky. Fear can keep people from taking necessary risks.

We have had to find a balance in motivating people to stay with us on our TQM journey. For some, just providing an opportunity to solve process problems has been enough. For others, the potential to save money is the key motivator. For others still, appealing to their sense of pride or need for prestige works. And, yes, we sometimes have to resort to fear. In general, we try to keep motivators as positive as possible, use more than one with any given individual, and keep in mind that sometimes nothing will work.

I have to admit, too, that my own enthusiasm has waned now and then. Other business demands did not always permit me to spend as much time educating and promoting TQM to others as I would have liked. However, something always happened to get me back on course. Our production manager would quietly tell me I had stopped listening to others. Or, a contractor would tell me the same thing (but usually not so quietly). The most significant lesson for me was learning to be patient. I eventually realized all of our problems with building homes are not going to be solved in one day.

Does Everyone Follow the Same Itinerary?

I have learned to accept that the many contractors and suppliers who work with Cannon Development do not all have the same understanding and commitment to TQM. But you don't need to keep everyone at the same point on the itinerary. Just try to keep them close to the same page. The latter task is far less difficult if you stay focused on core principles.

I do not demand buy-in to a vast number of TQM principles, tenets, or "rules." For me, the essentials boil down to only two—listen to your customers (including each other) and concentrate on continuous improvement. By adhering to these two tenets and by learning from our mistakes, everyone eventually comes to the same point. But because TQM is such a long-term effort we have found it unnecessary to micromanage too much. It's OK if everyone is not on the exact same paragraph as long as they want to be involved with us and believe in our two basic tenets.

Empowerment

On the surface, empowering the people who touch a process to make decisions about how to improve that process sounds like a tremendous idea. And it is. But internal customers, especially contractors, may have been beaten on for the greater portion of their business life. Some are reluctant to take on the responsibility and accountability that comes with empowerment. Others may have

---

Tip

TQM's Beauty Is Its Simplicity

I used to think TQM was a very complicated management practice. And while I suspect some who have taken this journey have indeed made it a complicated endeavor, we choose to keep things simple. We only hold to two tenets—listen to customers and improve processes. This makes it easy to communicate our intentions to those we invite to go along with us. It also helps keep us from getting distracted.

low self-esteem or low tolerance for criticism. After all, why would someone be willing to take responsibility for something they were punished for in the past?

Before empowerment can take place, people must first realize that mistakes will not jeopardize their jobs. It helps to recognize when feelings of self-worth are low and fear of criticism is high. An effective counter to such negative feelings is to show genuine interest not only in the work people do but in their personal views on life as well. I try to establish a common bond with each individual. If a genuine bond develops, the individual will be more willing to accept empowerment.

But there is a flip side to empowerment. Just how much decision-making power am I, the builder, to give up? At first blush, the answer is to give up all the decision making power other people are willing to accept, as long as they accept the accompanying responsibility and accountability. However, I tend to be a "control person." Too often I think I know more about the home building process than those who work around me. And, after all, it's my money.

It has been difficult for me to give up control over decisions, particularly from a scheduling or financial standpoint. But for TQM to work for us, I realized early on, this attitude had to change. I couldn't expect people around me to change their work attitudes if I did not change mine. Both the receivers and the givers of decision-making power have to feel comfortable in the exchange.

There can be people who seek power for the wrong reasons. Some people may want to use their new-found power "to show the world" or redress wrongs that they feel have been imposed upon them. But people with this bent may not be willing to take responsibility for their decisions even if (or when) the results are disastrous for customers.

In brief, my feeling on empowerment is this: I can put fuel in people's tanks and check the air in their tires. But they must be willing to turn the ignition and take the wheel. I can't make the trip for other people.

What Happens When the Market Goes Sour?

A serious test of our will to continue the TQM journey began in 1994. Housing starts in our area were down 25 percent that year. At the close of 1995 starts had dropped another 39 percent. While Cannon's starts were not down this much, our annual net profits during this time had dropped 22 percent.

Most readers, especially builders, will understand the impact a 54 percent reduction in market can have on business attitudes. Such a dramatic downshift gives rise to a lot of reassessment. But while I do not have a way to verify my own conclusion, I believe our TQM efforts since 1989 have contributed significantly to our ability to weather this storm in our local home building market.

To date, labor trades have suffered the most. Without retained earnings to fall back on, personal debt has piled up for some, and cash flows have sometimes run from break-even to negative. For our construction quality to stay high, we need these trades to survive. They have shrunk crew sizes and managers and owners have put their own toolbelts back on. Through all this, they have remained focused on continuous improvement.

At Cannon Development, it has been tempting to stop "listening and acting" in this survival atmosphere. We gave up the annual picnic this year to hold down expenses. We also cut back on production personnel and office staff work hours. But it seems unnatural for us not to continue daily to think in terms of what it is the customers want and what we can do to meet or exceed those expectations.

The attitude improvement we found with TQM has been more powerful than the significant setbacks we have faced. TQM has shown us the importance of maintaining a focus on *both* customer satisfaction and cost containment. And this outlook will see us through this difficult time—which some local economists believe will continue for at least three more years. However, had we started our journey in 1993 or 1994 rather than 1989, we would have had to work much harder to get where we are today. To be honest, I'm not sure we would have tried. In retrospect, this underscores how important it was for us to check our business environment before we began our journey.

What's Left to Do?

I have a strong sense we are making headway on our journey. When we see examples of increased buyer satisfaction, we take pride and personal satisfaction in the work we do. I am delighted by the large number of contractors and suppliers who have joined us on our journey. Not everyone committed in the early days, but most are on board now. In fact, several of our suppliers are now engaged in their own version of TQM. Most of the manufacturers whose products are handled locally by our distributors are also involved with TQM. Many have been involved longer than Cannon Development.

A recent experience with a framing lumber supplier really brought home how pleasant TQM can be for the external customer. One of our framers came to me looking for a price increase to help defray the expense of the considerable time he had to spend sorting through 2x10 floor joists looking for uniformity in width. He wanted to make sure the joists laid down did not have wide variance in width. This would help eliminate subsequent squeaks caused when plywood pulled away from low joists. This tradesperson was trying to do a good job as well as trying to help us avoid future problems. We were hindering him by not making sure he had the right material.

Realizing we both knew the real problem (squeaks) and the root cause (varying widths of 2x10 joists), we had our framers start measuring 2x10s received from different sources. We set the tolerance at 9-3/8" to 9-7/8", plotted the results on a simple run chart, and then passed the results along to the lumber distributor, who in turn passed the results to the plant supplying the 2x10s.

Within a week of turning the results over to the distributor, I got a call from a foreman at a lumber mill in Idaho. He informed me they were already working to solve the problem that caused so much variation in the lumber they produced. It felt really good to be on the receiving end of TQM.

One of the most pleasant results I have experienced from TQM to date was receiving our National Housing Quality Award in 1993. Being recognized by people in the industry also was a big deal for our contractors, suppliers, and employees. We have put in many hours to improve our ability to satisfy homeowners' needs. But we can still find ways to improve how we build homes.

Just how far do we have yet to go on our journey? And are our efforts really worth the time and expense we continue to invest? Evaluating our TQM effort provides some answers to these questions. How we do this, and what we have learned, is explained in the next chapter.

Chapter Five

Are We There Yet?

The last four chapters described in some detail what life can be like for the builder on a TQM journey. But our story is not yet complete. How do we know we are making progress? And how do we know we're on the right road? Had we waited for six years to answer these two questions, what we learned along the way would probably have gone for naught. If we are not making progress or are headed in the wrong direction, we need to make immediate adjustments. However, since there was so much to say about life on the road in general, I thought it best to wait until now to assess our progress for the reader.

Here is the rest of our story about life on the road.

Assessing Progress

Our sights are set on customer satisfaction and cost containment. How far we have traveled is judged against these two milestones. While a written mission statement, an active steering committee, and productive ad hoc teams are all necessary parts of TQM, the measures of real progress are satisfied customers and costs that are under control. We need ways to assess progress toward these goals.

Evaluation Plan

Since the very first days of our journey we have sought to assess our progress using a straightforward five-step method:

Step 1: Set goals for customer satisfaction and cost containment.
Step 2: Identify quantifiable indicators for each goal.
Step 3: Collect data on a set schedule for each indicator.
Step 4: Analyze the data and prepare summary reports of the outcomes.
Step 5: Compare each outcome with the goal and note yearly trends.

When a goal is met, we set a new goal. Likewise, if the yearly trend is positive, we continue along the same path. But if the trend is negative, we find out what we are doing wrong and take corrective action. If this doesn't work, we try something else.

We find the most effective way to collect information on customer satisfaction depends on the customer. Formal techniques such as surveys work best with buyers. More informal techniques such as brown bag lunches and gab sessions work well with our employees. Even more informal techniques, such as "TQMing," work best with contractors and suppliers. The most effective way to collect information on cost containment is to organize the data you already have about schedules and cost variances.

Evaluating Buyer Satisfaction

We have used a written survey to measure buyer satisfaction since the mid-1980s. While this has been an effective way to assess the quality of our products and work, we have improved the survey by changing both the number of questions we ask and the rating scale used by respondents. And as we discussed previously, we now use two surveys. We send a written survey to owners approximately six months after closing. The second survey, which I conduct by telephone approximately two months after move-in, has been in use for just over one year. Here is a chronology of the changes we have made and what we have learned about buyer satisfaction since we began our journey in late 1989.

Results of the Post-Closing Survey. In 1989 and 1990 our post-closing survey asked owners to rate on a scale of 1 to 5 (a "5" being high) the performance of different products used in their homes as well as the performance of our service and production supervisors. We also asked new owners to note how likely they would be to recommend us to friends. Here is a summary of how well we stacked up in these areas during the first two years of our TQM journey.

	1989 Average Rating	1990 Average Rating
Product Performance	4.26	4.06
Range	(3.0–4.56)	(2.92–4.33)
Supervisors		
Service	4.65	4.39
Construction	4.49	4.53
Recommend Cannon to Buy From	4.88	4.67

At first we did not have targets established for these ratings. In fact we did not begin to tabulate the data we had compiled until mid-1991. At this time the steering committee decided we should review the surveys each year to better detect any trends.

Overall we were pleased with our 1989 and 1990 ratings, but we felt we could improve in certain areas. For example, concrete floors by far received the lowest rating both years (3.0 in 1989 and 2.92 in 1990). As we discussed in Chapter Two, cracks in basement floors led the steering committee to commission an ad hoc team that ultimately made recommendations for solving the problem.

After reviewing the 1990 results, we decided the number of products rated should be increased to twenty-three. In 1992 we added four questions about service to the survey, and increased the number of products rated to twenty-five. Our goal during both years was to maintain an average rating of "4" in all areas surveyed. Here's how we did:

	1991 Average Rating	1992 Average Rating
Product Performance	4.14	3.90
Range	(3.50–4.83)	(3.13–4.50)
Supervisors		
Service	4.50	5.00
Construction	4.80	5.00
Service Work	not rated	4.26
Recommend Cannon To Buy From	4.50	4.50

While these numbers were good, something still didn't seem right. It finally dawned on us that we were not really measuring customer satisfaction. We were measuring how buyers liked various aspects of our work, but we did not know how closely these ratings were related to the buyers' own expectations. We determined that a more accurate measure of satisfaction might be the degree to which the four important aspects of our homes—service, products, staff performance, and our reputation—met their expectations. We changed the rating scale.

Our long-term goal is for each characteristic listed on the survey to receive a "5" rating, which indicates expectations are consistently being exceeded. Our short-term goal is to achieve a "3," which indicates expectations are being met. In addition, any rating below "3" calls for investigation by Doug, as our production manager, or by myself. We also follow up on any negative comment found on a survey, regardless of the rating given to the characteristic in question. Results for 1993, 1994, and 1995 were as follows:

	1993 Average Rating	1994 Average Rating	1995 Average Rating
Product Performance	3.88	4.00	4.01
Range	(2.83–4.09)	(2.50–4.75)	(3.00–5.00)
Supervisors			
Service	4.11	4.00	5.00
Construction	4.33	4.33	5.00
Service Work	3.88	4.00	4.67
Range	(3.55–4.17)	(3.67–4.33)	(4.01–5.00)
Recommend Cannon	4.73	4.75	5.00

Tip

What Does "Customer Satisfaction" Mean?

It was a long time coming, but we now realize "satisfaction" does not always mean "like." We may not particularly like trucks—but they are necessary if one wants to haul large loads. We may prefer cars, but they can't haul large loads. Customer satisfaction is the same way. Customers may or may not like something. The point is whether or not that something meets or exceed their expectations.

On the average we are doing very well in all four areas. The range of product performance in 1993 and 1994 included numbers less than "3." Those low ranges (vinyl flooring and caulking) have been addressed through increasing the product specification. The resulting increases in our overall house cost were offset to some degree through switching brands in another area. This switch was made with the approval of our sales manager, who indicated buyers were more interested in price containment than brand-name recognition for this particular product. We're very pleased with our first "5.0" on "Recommend Cannon to Buy From." We don't want to get overly excited, though, as both our sales volume and our percentage of questionnaires returned were down from previous years. This is one of those areas where we need to understand the numbers that make up the statistics we use and not rely on just the final result.

I should make one additional comment about how we use data from the post-closing survey. When homeowners rate employee performance, they in effect perform a "performance appraisal" that covers the team effort. We think this is a very meaningful way to appraise work. In TQM, satisfying the customer is what work is all about. I'll come back to the topic of performance appraisals later.

Using a Telephone Survey. Beginning in late 1994 Barb and I began to telephone new owners approximately two months after closing to discuss their overall satisfaction with their buying experience. Initially I did not use a set protocol for these discussions. I would take notes, but the number and type of questions I asked were not consistent from call to call. This inconsistency made it difficult to give meaningful and concise summaries of these conversations to the steering committee. We decided a better approach would be for me to write responses on a form I kept in front of me while talking on the phone.

We are currently investigating the best way to summarize the individual responses. Because the questions we ask are open-ended, the responses tend to be subjective comments rather than easily quantified ratings. I anticipate using a set format will make it easier to prepare reports for the steering committee.

TQM has taught us we are selling both a product and a service to buyers. The product is the home, and the post-closing survey provides information on how well we did with the products and construction that went into the house. The service is all the education and assistance we devote to buyers during the time of purchase and construction. The telephone survey helps us judge how well this service met expectations.

Evaluating Internal Customer Satisfaction

I believe using a survey to determine our employee's level of satisfaction is somehow artificial. These are people with whom I work on a daily basis. If I cannot "read" their behavior and their attitudes, then the trust that is needed for true teamwork does not exist. Because our company is so small, it is not necessary to use written surveys to make this determination. However, employee satisfaction is important and I do make a conscious effort to listen and observe.

Indicators of Internal Customer Satisfaction. There are two indications that expectations of all our internal customers are being consistently met (or even exceeded). First, they tell me during informal conversations how they feel about working with Cannon Development. Second, we have extremely low turnover.

We have had to replace only one full-time Cannon employee in the past six years, and that was because the person moved away from the area. The person's departure had nothing to do with performance or not feeling part of the Cannon office team. Likewise, we have had to replace only a handful of the contractors and suppliers who work with us. In only one case did the change reflect performance problems.

On several occasions I have asked Chuck to verify my perceptions about internal customer satisfaction. He has interviewed employees one-on-one and has met with contractors and suppliers without any Cannon staff being present. His reports indicate customer satisfaction is seldom an issue in these discussions. When internal customers have wanted to talk about issues, they have focused on concerns with process improvement.

What About My Satisfaction? As owner of Cannon Development I also am an internal customer. Thus my expectations should be met—and hopefully exceeded. My primary expectation is that everyone who works with me, either directly as an employee or indirectly as a tradesperson, will follow two basic TQM tenets: (1) listen to buyers and to one another; and (2) constantly seek better ways of doing things.

People around me demonstrate daily that they have adopted the TQM way of thinking. The many examples noted throughout this book are evidence enough of my level of satisfaction.

Make no mistake: I don't always succeed in my efforts, nor is everyone working with me always a delighted traveler. None of us are consistently delighted with our work. Nonetheless, we are a team. As with any team, some members occasionally have opposing views. However, in a true team environment, people with different viewpoints can raise them without being criticized by others for doing so. In addition, every team member knows that should immediate corrective action not be possible, he or she will be given a full explanation. This includes an explanation of how I read the situation and why I may feel the opinions of others need to be taken into consideration before coming up with a solution.

An atmosphere of trust is the hallmark of internal customer satisfaction. The upshot of this is that formal performance appraisals are not used at Cannon Development. I do not formally appraise the performance of employees, nor do they formally appraise mine. This connects to something Deming tries to say about TQM. One of his well-known "14 points" is to eliminate performance appraisals and ratings

Evaluating Cost Containment Efforts

We look at "costs" in two ways. One way is to track the direct expenses involved in construction and service. Another is to consider time as money and keep track of how long it takes to build and service each home.

Construction Costs. One of our initial goals was to hold actual construction costs to no more than two percent over estimate. Since 1989 we have failed to meet this goal only once. In 1993 I made poor business decisions affecting two homes. In one situation we "bought" a job in an area where we were losing work. This was a big project and I did a poor job estimating on the front end. Change orders added to the problem. To make matters worse, we had terrible site conditions caused by poor weather from stake-out to completion. Poor weather also caused over-runs in the second home. But overall, as the following chart illustrates, we have done a good job in meeting our goals. The actual annual variance across all jobs is shown on the second line.

	1988	1989	1990	1991	1992	1993	1994	1995
Goal%	none	2	2	2	1	1	1	1
Actual%	not measured	(1.05)	.60	(.69)	(.15)	2.96	(.076)	(.11)

(Parentheses indicate actual was less than estimated.)

Doug and I track the variance between actual and estimated costs in each of the fifty construction phases for each job. An analysis of our cost variance report tells us where we are over estimate in particular phases. We identify the root causes of any negative variance and take any corrective action that may be called for. At the end of each year we compile average variances for each phase. This report alerts us to negative trends. When these are found, corrective action is taken.

The production department's goal for 1996 is to keep actual costs to within .9 percent of estimate. In 1997 and 1998 we hope to keep this figure within .85 percent and .80 percent respectively. Later in 1996 the steering committee and I will assess whether or not we should continue to set new goals in this area. I am not sure it is realistic to reduce cost variance any further. The additional effort required will not produce that much of an increase in profitability.

We have achieved considerable success at cost containment without compromising our quality standards. I don't want people to look for shortcuts that may save money but that increase our or the buyer's service costs after the warranty period. It would be possible in our part of the country, for example, to use a less expensive heating system. But while such a system will last beyond the manufacturers' warranties, it will have to be replaced sooner than the system we currently use.

We could save some money up front, but doing so would cost the owner more in the long run. My preference is to focus our efforts on *increasing* the standards we have set without proportionately increasing costs. The steering committee and production department continue to monitor our performance and employees also have an incentive through our quarterly bonus plan to see that we don't let things slip. We will continue to look for other ways to contain construction costs but we now are shifting our attention to increasing profitability through increased sales.

Service and Warranty Costs. Direct expenses associated with service and warranty repairs are accounted for in the overall cost of building the house. While we watch the actual expenses incurred here, and typically nothing appears out of line, the time it takes to complete the work has been our major concern. We believe our buyers don't want us around after closing, so the less time we spend in their homes the better. Spending less time on callbacks also makes our contractors and suppliers happier. The upshot is this: We are less concerned about the number of problems than we are about the time it takes to correct a problem and whether a particular problem persists.

In 1988, before we started our TQM journey, it took us on average more than 48 days to complete service work. The production department recommended that our initial target should be set at 28 days.

By the end of the first full year on our journey it appeared this was an attainable goal. So the department decided to reduce the target each year over the next five years. Here is a summary of what happened:

Year	1988	1989	1990	1991	1992	1993	1994	1995
Goal (days)	none	28	28	21	16	12	9	7
Actual	48.6	35.1	30.5	24.2	17.8	12.2	9.4	4.2

The production department has set even tighter goals for 1996 through 1998 (6, 5, and 4.5 days respectively). Once again, however, I am not sure these are realistic figures. The steering committee will be taking another look at this goal in the coming year.

We also sought to reduce the time required to complete annual service. In 1988, it took us on average more than 37 days to compete this work. Again, the production department felt a reasonable target would be 28 days. And though we failed to meet this goal in 1989 and 1990, the department still wanted to set tighter goals over succeeding years. Here are the results:

Year	1988	1989	1990	1991	1992	1993	1994	1995
Goal (days)	none	28	28	21	16	12	24	17
Actual	37.1	56.9	36.7	21.5	37.0	45.6	27.8	17.2

As you can see, service over-runs blew us off the road in 1989, 1992, and again in 1993. While one may excuse 1989 since we didn't start TQM until late that year, we knew at the time we had a problem with completing annual service. However, the causes behind the problems we experienced the other two years were clear.

In both 1992 and 1993 we assigned a new person to oversee completion of annual service. This person also had a long list of other responsibilities, and the production manager and I failed to emphasize enough the importance of timely completion of service work. We also failed to properly train these two individuals and follow up on what they were doing. This clearly was management's fault. We could not blame the buyers, the tradespeople, or our employees for not meeting our goals. But another dynamic affecting annual service merits discussion.

Our buyers tend *not* to be too concerned about having their annual service work completed swiftly. Drywall repairs are the most common callback we have at the year-end review. Yet we have a devil of a time finding a time for the drywall contractor and the painter who follows him to get into the homes to make these repairs. Both contractors are quite agreeable to set aside the time to do this; but homeowners don't seem to mind if it takes the contractors several weeks to find a convenient match with their schedules.

Our homeowners have become more casual about the timing of service work than Cannon Development and our contractors. Buyers have come to understand that if we say we will take care of an issue, we will. And the types of annual service items required apparently are a low enough priority that owners value convenience over speed in scheduling the work. Our standard for competing service work is therefore self-imposed, not buyer-imposed; and the production department has set this standard at 14 days for each of the next three years. Reducing this goal to less than 14 days may not be justified based on customer expectations.

Cost Savings from Reduced Cycle Time. We recently added a new goal. The cycle time for our homes varies with the size of the home we build. Annually, across all jobs, we come in within one day of the estimate. Doug and I believe there is room for improvement, however. In fact, we think this variance can be held to zero days. Let me explain how we plan to do this.

Our long-range schedules show variance in estimated versus actual time required to complete each construction phase. When we find an actual time greater than estimated we consider what, if anything, can be done to correct the situation. The solution may an adjustment in our specifications, or there may be procedural improvements that could help us hold to the original time estimate.

In the coming year, two ad hoc teams of contractors and Cannon Development production employees will investigate ways we might increase the number of square feet we build per day. Currently this figure is approximately 20 square feet per day. One team will focus on what we do up through drywall. The second group will look at what we do from drywall to completion.

But I am not interested in reducing cycle time per se. I want to reduce cycle time while maintaining or even improving our construction standards. This is a point that I will stress when we formally commission the two ad hoc teams. But I have no doubt they are up to this challenge. And should the teams be successful, we will share part of our cost savings with those contractors who succeed in reducing cycle time in their phase of construction while meeting or exceeding the agreed-upon standards.

Assessing Our Progress

Our approach to assessing how TQM principles and techniques work for us has been disciplined but simple. We try various things and watch what happens. If things work, we keep doing them. When something indicates a problem, we make adjustments. If the adjustments don't work, we try something new. But don't get me wrong: it takes discipline to keep at this.

Assessing the Road Map

Early on, we spent considerable time mapping out how we wanted to pursue customer satisfaction and cost containment. This road map is embodied in our mission statement. The original multi-page statement explained the importance we placed on customer satisfaction and cost containment as our measures of a "quality home." The document also identified the principles we were going to follow in our pursuit of quality. We wanted everyone working with us to know up front what we meant by quality. The original document then went on to list all the business practices we felt were important to completing our mission.

For the first time, we had something in writing that represented what Cannon Development owners and employees stood for. This is why I have referred to TQM so many times in this book as our "way of doing business."

In the past we thought quality was only something important to buyers. And production goals, especially cost containment, were only important to internal customers. Now our notion of "customer" includes both buyers and internal customers, and "production" is not seen as something distinct from "quality." It would have been difficult to integrate TQM and business practices had we not changed our way of thinking.

Many of our contractors and suppliers, upon seeing our quality mission statement for the first time, told me they now understood what we were trying to achieve and how we were going to do it. More importantly, they knew what I was asking them to get into if they chose to go along with us.

Over time, the format of our mission statement has been refined. We found a shorter version, such as the one reproduced in Chapter One, was all anyone needed in order to grasp what we were doing. Eventually, we realized even this shortened version was no longer needed.

The message had been internalized and no one was asking to see the mission statement. Buyers and others outside our organization (such as real estate agents) were no longer inquiring about our TQM effort, let alone our mission. Most already knew we were "into TQM." So we refined the format even further. All we use now is a brief summary of the mission printed on the backs of our business cards. The statement reads:

OUR MISSION

Our mission is to provide the highest quality homes and services to our customers.

Quality means using products which provide our customers the best value with the highest degree of workmanship and expeditiously supplied services to meet or exceed their expectations.

These changes in format represent progress as we have internalized our sense of our quality mission. What is more significant, however, is that we have not found it necessary to change the message itself.

Assessing the Use of Teamwork

The steering committee has been an important "wheel" in our TQM journey. This team has been active since the early days in overseeing our efforts. But teamwork has taken on a much broader meaning at Cannon Development.

Company Employees. Cannon employees now function as a team. This is something I hoped would happen. I have only had to make one major adjustment in the way we conduct business to ensure we continue to work together in this fashion. Since we started in TQM, production employees have had a higher degree of involvement in setting production goals, and of course they all have been actively involved with the steering committee and ad hoc teams. But only a few were involved with annual operations planning. Now all employees participate in planning. This represented quite a change in our office culture.

Ad Hoc Teams. Ad hoc teams have also played an important role in our TQM effort. These groups have come up with several significant improvements to the way we build homes, and they have been instrumental in cementing the notion of teamwork throughout our extended organization. Ann Moffit, our office accountant, sums it up best. She says her work on the ad hoc teams has helped her feel more involved with our business. And it most certainly has opened her eyes to how difficult it is to reduce costs— both in time and in money. She also finds the reporting mechanism we use with the steering committee serves as a control over jumping to conclusions without thorough investigation.

An Evolution in the Concept of Teamwork. Ann's opinion, which I share, is that a sense of teamwork is much more prevalent today than it was in 1989. Daily working relationships in the office and on the jobsites unfold in a more natural, informal way. When confronted with a problem, the first thought is what can "you and I" do to correct the situation, not "Is this something that needs to go to the steering committee?" We still use formally commissioned ad hoc teams, but not as often as we did in the early days of the journey.

Assessing Our Approach to Problem Solving

Assessing how we solve problems is really an assessment of how we make decisions. Many books on quality talk about formal six- or eight-step decision-making models; and several describe the "Plan-Act-Measure-Evaluate" approach. However, we find such formal models don't work for us all the time, especially during those hectic moments out on the jobsite when only one or few people are involved. Formal models do help when there are competing resources and many people interacting with you.

I would like to offer the following seven points based on what I have learned about problem solving and decision making:

1. Solving any problem is much easier to do when *we* work on it. The combined horsepower of two or more cylinders is much greater than one cylinder working alone.
2. *First* find out what's wrong and what causes it, *then* fix it.
3. Assume everything is broken and will always need fixing, but don't try to fix everything at once. Start by listening to customers.
4. Learn to recognize when others are just blowing off steam and when they are asking for help.
5. Don't throw the baby (customer satisfaction) out with the bath water (formal TQM).
6. When something doesn't work, try something else.
7. Learn to take on only what the weakest link in the organization can handle and work on strengthening the weakest link.

Assessing the Cost of TQM

During our first year in TQM, Cannon Development spent approximately $3,000 on training and consulting. Most of this was spent during the planning phase. Since that time, very little money has been spent on training. Because Chuck donated some of his consulting time, our consulting fees were kept fairly low. We realize that

Tip

A Good Way to Evaluate Your TQM Effort

I think it is a good idea for a company pursuing TQM to periodically have an objective, outside assessment of its pursuit. However, such an assessment can be expensive. Consulting fees alone can easily exceed $2,000 dollars per day—and one day is probably not enough. There may also be travel-related expenses.

We found the application process for the NHQ Award an excellent way to evaluate our progress in TQM. This competition not only spurs rigorous self-evaluation, but provides a mechanism to receive comments and insights on a TQM effort by objective outside observers. There are some direct expenses involved. For example, you may need a consultant to review the results of your self-assessment; and there is an application fee. But these expenses are still much less than what you would pay to have a full, formal assessment conducted by a consultant.

other companies will not be this fortunate; nonetheless, if the quality consultant does a good job during the early days, a company should be able to handle most future training on its own. We only find it necessary to consult with Chuck once every twelve to eighteen months, and for only 1 to 2 hours each time he comes in. Such sessions typically cost us less than $250.

The biggest cost associated with TQM is time, especially in the early days. We estimate the eight Cannon employees and I collectively spent over 500 hours during the planning phase. And this time was devoted just to deciding if we wanted to make the journey!

But remember, if TQM is to work it must become an integral part of daily business functions. We have not allowed any of our teams—to spend meeting time shooting the breeze or pointing fingers. Teams are expected to focus on process improvement; and that's exactly what they do. Time spent learning to do things right the first time saves the hours of time it takes to correct mistakes.

Once we were underway, we found the costs of TQM diminished considerably while the benefits continued to grow. This is what I think Philip Crosby means when he says "Quality is Free."

As time goes by, we expect to get a better handle on how much money we lose to nonconformance (see *Principles of Quality Costs* in the resource list). I sense we are doing pretty well. Our cost containment goals are continually met, our indicators of customer satisfaction are positive. But to be honest, I cannot put a dollar figure on how much nonconformance is costing us and I think it makes prudent business sense to be able to do so. By the fall of 1996 we plan to install new computing equipment and software that should help us better track these costs.

Assessing Competitiveness Through Benchmarking

To borrow a phrase from Mark Twain, whenever people think they are on the right road, they still need to keep an eye out for the other cars. The best way I know to do this is benchmarking, which is the practice of comparing your business practices to those used by companies considered to be industry leaders (see *Benchmarking* in the resource list). Looking at what other builders do provides a basis

> **Tip**
>
> **Finding Benchmarks**
>
> The companies we choose to benchmark come from both within and outside the home building industry. We look to those builders in our locale who have a reputation for being good at what they do. If you don't know who those builders are in your locale, all you have to do is ask around. People will tell you. As for companies outside the industry, we have found organizations such as Lexus automobile dealerships excellent sources of good ideas, especially for ways to measure buyer satisfaction.

for confirming that we are on the right road. And we often find ideas that spark our own creative thinking.

Benchmarking Administrative Practices. I often wonder whether or not the myriad administrative practices we follow are really the best way to do things around the office. For example, we provide extensive benefits for our employees. We also have a bonus plan. And, as is probably obvious by now, we collect a good deal of information and data. I believe all such practices are necessary and appropriate. But what do others do?

I am fortunate to participate in a small, informal network of local builders who are highly regarded within the local home building industry. They are also willing to share their thoughts with me on business practices. While this process is highly subjective, I find this an effective way to benchmark how we run our business.

Benchmarking Construction Practices. Our construction practices are codified in the specification manual. While I do not carry this manual around, I do informally benchmark these practices in two ways. When I talk with contractors and suppliers who work for other builders, I often find they are willing to share perspectives on construction processes and techniques they see "around town."

Another way to benchmark construction practices is to attend local home builders' association meetings and social events. Many informal discussions take place at such gatherings. This is an excellent place to learn about other builders' practices and vendors' new products.

Benchmarking Marketing Practices. The evaluation of customer satisfaction and cost containment I described earlier provides a basis on which to judge the effectiveness of our house designs. We have found another way to judge design is to compare our features with those offered by several local builders. We also benchmark the prices we charge for our homes against what the other builders charge for homes with similar features.

We have used the "secret shopper" approach from time to time; at other times, one of us just walks in and tells the sales person why we are there. I'm not sure whether other builders shop us, but I hope they do. In any event we learn a great deal about our competitiveness by shopping the competition.

What Have We Learned?

Chuck Layne and I have spent many hours talking about how and why Cannon Development successfully implemented TQM. Many others have tried and failed. Almost daily we see newspaper and professional journal articles that chronicle such failures and false starts. While Chuck and I don't have definitive answers as to why others fail, we do have some thoughts which may explain why TQM is alive and well at Cannon Development even in our current trying times.

Misconceptions about TQM

There are several commonly held ideas about quality improvement that don't hold up. Chuck and I call them myths. Being aware of these common misunderstandings may help you avoid some difficulties on your TQM journey. It certainly helped us.

Myth: TQM Solves All Business Problems. The reality is that TQM is not a panacea for the numerous negative legislative, societal, and economic conditions facing our business. If you are searching for one, TQM is not your answer. In addition, TQM is not a quick fix to *any* negative business condition. TQM takes time—which is something you may not have. Given such realities, what business problems can we expect TQM to solve?

For one, we can expect *continuous improvement* to the different phases of home building as buyer and internal customer expectations become more quantitative and communication improves. Even here, success comes in bits and pieces; and it is uneven throughout the network of internal customers involved with building a home. In other words, continuous improvement really means aggregate improvement. It most certainly does not mean constant or universal improvement. There just haven't been enough significant breakthroughs in our processes to move things along more quickly.

We can expect reductions in operating costs when we reduce rework and waste. We can also expect increases in sales when customer satisfaction indices improve and become public knowledge. While cost containment alone is insufficient to sustain a healthy bottom line, the combination of decreased costs and increased sales is a strong one.

Over the long haul some of us will survive the many external threats to our businesses. At least following the tenets of TQM helps keep us in business long enough to try. If we fail, it won't be because we didn't take action on what we learned from listening to customers.

Myth: TQM Requires Lots of Training. Training *is* required to properly undertake a TQM journey. People need a formal orientation to TQM. And, as illustrated earlier, buyers generally need to be educated about the homebuilding process. But regardless of who needs to be trained (or educated) and what they need to learn, training should not be provided in large doses. People need time to absorb new concepts and to apply what they learn. Training and other formal educational activities, like material arrival on the job, is best provided "just in time."

The key here is to think in terms of learning. Terms such as "training" or "education" connote special activities reserved for a specific place and time, and led by someone we call a trainer or facilitator. However, learning takes place in many places, and hopefully it takes place all the time. A conversation on the jobsite, a review of survey information, an ad hoc team meeting, and "TQMing" are excellent examples of where we learn about customer satisfaction and cost containment.

A second key is to remember that internal customers, steering committee members, ad hoc team members, and even business owners and senior managers will not learn anything if they don't want to. Throughout our journey I have tried to involve everyone in our extended organization, provide them with support and recognition, communicate in a variety of ways, and remain sensitive to the things that keep people motivated. This is hard work but it is critical to success with TQM. That's why I devoted so much of the previous chapter to this topic.

TQM requires lots of learning. And experiencing customer satisfaction and cost containment firsthand is a great teacher. If something works, keep at it. If it doesn't work, try something else. Learn from your mistakes *and* your successes. And keep trying. This is what the word "continuous" means in the phrase "continuous improvement."

Myth: A "Quality Builder" Is One Who Already Knows How to Build Better Homes. In reality it is only through listening to buyers and internal customers that we learn how to build a better home. This gives us the information we need to find products or adjust processes to match the customers' desires. It also gives us the information we need to educate when necessary so that customers understand why what we are offering is indeed better. A "better home" ultimately is defined by those who buy the home.

In the end, true buyer satisfaction will only be achieved if you produce a house that is of high quality. A house with deficiencies in design, materials, products, or workmanship will not satisfy buyers six months or six years after closing even if it looked great during the final walk-through. And it does little good to meet any customer's expectations if employees, contractors, suppliers, and buyers are frustrated and generally worn down when the house is completed. That is why TQM places so much emphasis on listening and being responsive to customer needs, on communication, on teamwork, and on cost containment.

We always strive to keep in mind the customers' point of view. The building experience should be pleasant all around, and we don't want internal and external customers to go mentally or financially broke in the process.

Myth: TQM Means "Zero Defects" or "Getting it Right the First Time." We do not live in a perfect world. Humans, not machines, build homes. An extended network of many people using relatively unstable materials in an unpredictable physical environment build homes. On top of this, we continually encounter changing buyer expectations. These expectations are getting higher and each buyer has a set of expectations that is slightly different from every other buyer. Warranty standards past buyers accepted as "the" standard for our homes are now seen by most as the "minimum" standard.

These variables make it impossible to prevent all confusion and problems. But we should still strive to minimize the variability in homebuilding. Buyer education and clearly written, detailed contractual documents will help. Having all internal customers help us write, update, and follow material and labor specifications also helps. And striving to build a home that meets a customer's expectations rather than mine (or an architect's) helps. While our goal is to achieve as close to zero defects as humanly possible, eliminating all variability is unrealistic.

The phrase, "Getting it right the first time" sounds great but does not mean perfection. If we spend enough time in planning, most times we will "get it right the first time." When it doesn't, we want to make sure we don't make the same mistake twice. In other words, we learn to get it right next time.

Myth: TQM Requires the Use of "Statistical Process Control." One commonly held TQM principle that is often misunderstood (and misused) is that solving process related problems requires knowing how to collect, analyze and interpret data. Collectively these techniques are often referred to as "statistical process control" or SPC. The word that seems to get people in trouble is "statistical."

One very basic definition of statistics is "using data to make decisions." How we at Cannon use data to make decisions has been discussed at length. But the kinds of data we use are not "statistical" in the conventional sense. We don't use Pareto charts, standard deviations, process capability indices, and control charts. For most of us, at least for me, that level of statistics is pretty frightening stuff. But more importantly, I have seldom if ever found these statistical methods useful. Chuck, who is well versed in SPC, has spent several hours working with me, our steering committee, and ad hoc teams investigating how we might be able to use this higher form of statistics. So far, however, we just haven't found it that useful for our operation.

My attention is on data that tell us something we need to know. Simple percentages, averages, and graphs satisfy the greatest amount of our needs. While it is perfectly acceptable, I just do not choose to call these techniques SPC. And I feel no guilt for not using standard deviations, histograms, and control charts.

Myth: TQM Means Achieving Consensus on What Satisfies Customers. The people who work with us do not always agree on what should be done to satisfy customers. Sometimes there are competing ideas; sometimes people just don't know (or care). However, when an issue lacks consensus, someone still has to make the decision on what to do.

For a builder, TQM decision-making is not always rooted in democracy. While consensus-building is fine, the final decision on what to do to satisfy customers falls on the shoulders of the builder. The only problem I have struggled with in this regard is knowing when to shut down discussion and move to a decision. Being good at this appears to be somewhat more art than science.

Myth: We Must do Everything The Quality Gurus Tell Us. Organizations like ours that have been involved with TQM for several years are finding not all of what the architects of TQM say works exactly as presented. For example, Deming says there are 14 principles (often referred to as "Demings 14 Points") that must be followed if TQM is to be successful. Crosby also has 14 principles to follow. Some are clearly the same as Deming's and some appear to be different. But to be fair to Deming, Crosby, and other lenders of this movement, there is nothing wrong or detrimental in what they all have to say. The problem is in how their ideas are interpreted by the populace.

As with any widespread management strategy, many popular notions about TQM go beyond what its proponents have espoused. I believe too many people have gotten caught up in philosophy and technique, and have lost sight of why they became involved in TQM in the first place.

TQM is merely one way to improve a business. You may or may not find that all TQM techniques and principles apply to your situation. For example, supplier certification and reengineering have not been covered in this book because we at Cannon have not made them central to our TQM effort. Other businesses do. Given time, you'll find the ones that serve you best. However, if your business is in serious financial trouble, you probably won't have time to investigate let alone implement TQM.

You can expect to experience frustration and anxiety when your circumstances don't seem to fit with anything you read about TQM (including this book). But don't let this deter you. As the manager leading your company in TQM, you will be called upon to make judgments on what to do when circumstances arise that are not covered by what you read or hear about TQM. Once again, let experience be your guide. You have gotten to where you are today by being right at least 51 percent of the time. TQM just helps you improve this percentage.

What Have We Not Considered?

In 1989 I thought TQM was just for big corporations. Chuck Layne told me it could be used by any size company. He also told me we could learn from the experiences of organizations many times our size. And he was right on both counts. In 1989 I also thought TQM meant following a lot of rules, using lots of high-powered statistics, and running a business by committee. Now I realize that TQM, for us, really means having a shared disposition (or attitude) to listen to all customers and each other and to take action on what we learn.

The way we implemented TQM occurred in a context that may be different from what other builders have to contend with. For example, I'm sure a Virginia builder deals with different buyer expectations for how a house should "weather" snow and extreme cold than we do in western New York. I also suspect that New York contractors may be more willing to work in severe winter weather simply because such weather is more of a fixture in their business environment. The specific problems addressed by builders in each area will differ, as will the solutions.

In some locales contractors are unionized, while in other areas, like ours, they are not. From what Chuck tells me about other industries, TQM can work in a union environment—but only when the union is fully involved with the effort. There have been instances in which unions have been the first to propose the consideration of TQM for a particular industry. Being unionized does not mean you cannot use TQM. However, unionization will likely affect the way you go about preparing for the journey. The bottom line is that the *why* behind what we do to implement TQM is more important that *how* we do it

Just How Far Have We Come?

After being on this journey for almost seven years, Cannon Development's understanding of quality and TQM has matured. We've moved from a lot of guessing and reacting with gut feelings to a lot of listening, measuring, and acting on the information our internal and external customers give us. But we have not yet reached the "end" of our journey. All of our customers are not satisfied and all costs are not completely under control. We also see the need to do more benchmarking of other builders' business practices in order to better assess how well we are doing. What I personally have come to recognize is the value of people in running a business. It's not about progress, flow charts, or written procedures. It is about personal credibility and uncompromising integrity in working with others. The quality process cannot start until there is a respect for the needs of others and a desire to solve their problems in an unobtrusive way. By understanding this simple principle, any builder can design, build, and deliver a home that will satisfy and delight the buyer.

When you purchased this book, you became our customer. How well we have met your expectations will be determined by what you do next. If we've provided some incentive for you to give TQM a try, or strengthened your commitment to your current TQM efforts, or provided you with some benchmarks, or given you some tips on how to smooth out your path, then in a sense we've accomplished what we set out to do. We hope the money and time you spent with this book have at least met your expectations.

Resources

General References

The references listed below typically do not include examples from the home building industry. We have found many of the references useful, nevertheless, in gaining a better insight into TQM and the quality movement in general. Take from these resources what makes sense to you. Try some of the ideas you like. If they work for you, keep using them. If they don't, try something else. One final word of caution: Some of the books we have listed may come across as "cookbooks" for quality improvement. I don't believe an adequate cookbook exists. I recommend you read them to better understand notions of customer satisfaction and cost containment, and for some ideas you might like to try.

Books

A Better Idea: Redefining the Way American Companies Work, by Donald Peterson and John Hillkirk (Boston: Houghton Mifflin Company, 1991). This text chronicles how Peterson, former CEO at Ford Motors, used many of Deming's notions about TQM to make significant improvement to customer satisfaction and cost containment at Ford Motor Company.

All I Really Need to Know I Learned from Watching Star Trek, by Dave Marianccio (New York: Crown Publishing Company, 1994). A very entertaining look at how this famous TV series illustrates both the positives and negatives of contemporary business management practices. Those familiar with the series will particularly appreciate the author's thoughts on corporate mission statements.

Benchmarking: The Search for Industry Best Practices that Lead to Superior Performance, by Robert C. Camp (Milwaukee: ASQC Press, 1989). ASQC stands for American Society for Quality Control. Many people consider Camp's book to be the authority on benchmarking.

Corporate Philosophies and Mission Statements, by Thomas A. Falsey (New York: Quorum Books, 1989). Falsey not only illustrates many different forms of mission statements but also does a good job in explaining all the purposes a mission statement serves.

Cultural Shift: The Employee Handbook for Changing Corporate Culture, by Price Pritchett (Dallas: Prichett Publishing Company. 1993). This work is thin on theory and full of ideas one can try to change an organization's culture. I found many ideas helped clarify some of my own experience and thoughts.

Dr. Deming: The American Who Taught the Japanese about Quality, by Rafael Aguayo (New York: Simon and Schuster, 1991).

How to Implement Total Quality Management, by Miles and Donna Southworth (Livonia, NY: Graphic Arts Publishing, 1992). This book cuts out a lot of jargon and preaching on the theory of TQM. It presents a very nuts-and-bolts approach to TQM.

KAIZEN: The Key to Japan's Competitive Success, by Masaaki Imai. (New York: Random House, 1986). This book helped explain to American business managers and owners what it means, from the Japanese perspective, to satisfy customers. This book helped me understand the difference between pleasing customers and placating them.

Let's Talk Quality, by Philip Crosby (New York: McGraw-Hill Publishing Company, 1989). This book, a follow-up to the author's earlier best-seller *Quality is Free* (see below), explains in great detail his 14 points of quality improvement.

Principles of Quality Costs (second edition), edited by Jack Campanella (Milwaukee: ASQC Quality Press,1990). This book is an excellent primer, full of examples from different types of companies, on how to track the costs and benefits of quality improvement efforts.

Quality is Free, by Philip Crosby (New York: McGraw-Hill Publishing Company, 1986). First published in the mid-1970s, this best-seller presents Crosby's argument on why companies should focus on customer satisfaction and cost containment. The money this makes a company more than outweighs the cost of the effort.

Quality Improvement Methods for Home Builders, by E. Lee Fisher (Upper Marlboro, MD: NAHB Research Center, 1995). This book contains a logical systems approach to the quality improvement process. Author Lee Fisher investigates the underlying causes of the quality problems involved with home building and discusses quality control methods and shows how they reduce construction costs while increasing quality and value.

Quality in America: How to Implement a Competitive Quality Program, by V. Daniel Hunt (Irwin Professional Publishing, 1991). Author V. Daniel Hunt analyzes the state of quality practices in America and offers his own plan for successfully implementing a competitive quality program based on the "Quality First" methodology. Hunt shows how to review quality performance and dramatically improve market share, profitability, and performance.

Quality or Else: The Revolution in World Business, by Lloyd Dobyns and Clare Crawford-Mason (Boston: Houghton Mifflin Company, 1991). This text is crammed full of explanations and detailed discussions on why American business must focus on customer satisfaction and cost containment in order to survive in a global economy. It also includes information on Deming's and Crosby's 14 Points.

Quality: The Myth and The Magic, by Cynthia Lane Westland (Milwaukee: ASQC Quality Press, 1990). This book extols the benefits of TQM and the pleasure one derives from being involved. It also looks at some myths similar to the one's addressed in Chapter Five.

Quality, Productivity, and Competitive Position, by W. Edwards Deming. (Cambridge, MA: Massachusetts Institute of Technology, 1982). One of the first books written by Deming that explains his approach to TQM in great detail.

Reengineering the Corporation: A Manifesto for Business Revolution, by Michael Hammer and James Champy (New York: HarperCollins Publishers, Inc., 1993). Hammer and Champy coined the phrase "re-engineering" with this publication. Included are extensive discussions of why one re-engineers, how it is done, and several case studies of those who have successfully re-engineered all or part of their organization.

The Deming Management Method, by Mary Walton (New York: Putnam Publishing Company, 1986). This is considered by many to be the best and most comprehensive work on Deming's approach to TQM.

The Fifth Discipline: Mastering the Five Practices of the Learning Organization, by Peter Senge (New York: Doubleday Publishing Company, 1990). This text makes for pretty difficult reading, but it does provide a sound background into why companies must focus on learning to be successful at whatever they want to do.

The Memory Jogger. (Methuen, MA: GOAL/QPC) Goal/QPC can be reached at 13 Branch Street. Methuen, MA 01844. GOAL/QPC is a non-profit organization devoted to promoting quality improvement. This inexpensive pocket-guide provides guidelines for organizing and interpreting data. Several examples of each technique are also provided.

What is Total Quality Control? The Japanese Way, by Kaoru Ishikawa. (Englewood Cliffs, NJ: Prentice-Hall, Inc., 1985). This book is for people who wish to understand the theory and application of SPC in Japan. This approach was taught to the Japanese by Deming, an American engineer, in the early 1950s.

Work Habits For a Radically Changing World, by Price Pritchett (Dallas: Pritchett Publishing Company, 1994). The style of this book is very similar to Pritchett's work on changing corporate culture noted above. Easy reading and many ideas worthy of consideration.

Your Attitude Is Showing: A Primer of Human Relations, by Elwood N. Chapman (New York: MacMillan Publishing Company, 1991). Written in a very informal, nontechnical way, this book offers many good ideas a manager can use to improve on-the-job communication and promote motivation of co-workers.

Reports

"How to Verify Training Required for Supplier Certification," by Charles A. Layne. In: *The Quality Control Scanner,* Volume 13, Number 9 (Livonia, NY: Graphic Arts Publishing Company). This article, which focuses on how to verify training requirements, illustrates what a company must do to become a certified supplier.

"Why the Fuss Over ISO 9000," by Charles A. Layne. In: *The Quality Control Scanner,* Volume 13, Number 11 (Livonia, NY: Graphic Arts Publishing Company). This article explain in some detail what ISO 9000 is about and how one achieves certification to the standards.

Best Practices Report: An Analysis of Management Practices that Impact On Performance, by Ernst & Young and The American Quality Foundation. (Milwaukee: ASQC Press, 1992). This is a report from a special task force that examined what worked and what did not work for several hundred companies that have successfully implemented TQM.

NAHB Research Center Report: Advanced Total Quality Management for Home Builders, by Edward Caldeira (Upper Marlboro, MD: NAHB Research Center, 1995). This report introduces a quality management model for home builders based upon the criteria of the National Housing Quality Award and the Malcolm Baldrige National Quality Award. The report presents an advanced examination of total quality principles, tools, and techniques.

Periodicals

Quality Progress, published by the American Society for Quality Control, 611 E. Wisconsin Avenue, PO Box 3005, Milwaukee, WI 53201-3005. This monthly journal devotes various issues to TQM, supplier certification, ISO standards, and other quality improvement approaches. ASQC also publishes an extensive list of books, videos and other documents related to quality. Some of these contain information geared specifically to the home building industry.

Quality Service Connection, published by the NAHB Research Center, 400 Prince George's Blvd., Upper Marlboro, MD 20772-8731. This newsletter is designed to help builders produce the highest quality construction and provide legendary customer service. It promotes Total Quality techniques geared toward customer satisfaction, long-term company growth, and continued profitability.

Training, published by Lakewood Publications, 50 S. Ninth St., Minneapolis, MN 55402. This is the most widely read monthly journal on issues related to training, education, and development in the workplace.

Services Available from the NAHB Research Center

More information on each of the following services can be obtained by calling the NRC at (301) 249-4000.

Builder's or Remodeler's Total Quality Survey. This survey is completed by your employees and analyzed by the NAHB Research Center. Results for the company are tabulated and compared to responses from other quality-minded builders or remodelers. Includes a valuable report of strengths and suggestions for continued improvement.

Builder's Total Quality On-Site Assessment. A National Housing Quality Award examiner performs one to two days of on-site interviews with management, employees, and the company's trade contractors to examine total quality practices. Detailed analysis produces a final report with findings and recommendations.

Total Quality Seminars. Designed for builders new to total quality, these seminars introduce builders to quality concepts such as quality culture, process improvement and problem solving, team building and partnering, and fact-based decision making. Seminars available to builders and home builders associations individually or as a series.

TQM for Small Business Video Workshop. A five-session video workshop by Shelley Hayner provides a place to start implementing TQM. Improve communication and build teamwork by inviting trade contractors to train with you as you learn. The result is improved customer satisfaction and profitability for everyone. A TQM Workbook to accompany the video is also available.

Do you have a story to tell?

We hope you enjoyed reading *Destination: Quality* and that this chronicle of a real builder's experience with TQM has provided useful ideas you will be able to apply in your own building business. Home Builder Press exists to publish practical business, construction, and other information for people who work in the residential building industry. On the back of this page you will find descriptions of a number of brochures many builders find helpful in educating their customers.

Most of our readers—and many of our authors—are builders, remodelers, and developers—businesspeople who are deeply involved with every aspect of running their companies. We welcome proposals for book or brochure ideas in the areas of land development, remodeling, business management, sales and marketing, residential construction and design, light commercial construction, multifamily, and seniors housing, including interesting case studies.

If there's a book or brochure you think we should publish—or if there's a book you'd like to write—please contact Home Builder Press at (800) 368-5242, ext. 222.

Helpful Products from Home Builder Press

Your New Home and How to Take Care of It

Something to give every customer at closing, or sooner. This beautifully designed 60-page-handbook, with its colorful cover, gives new homeowners tips about how to care for and maintain their homes. From air conditioning systems and appliances to water intake valves and windows, they will find simple, useful explanations in a quick-reference format, with a convenient "Owners Maintenance Record" and space for listing suppliers and contractors. $22 for set of 10.

The Design-Build Advantage

Sell your customers on the benefits of your firm's combined design and build services...and get more jobs! Equally applicable to new construction and remodeling design-build services, this brochure spells out the advantages of design-build to your potential or new clients. $15 for set of 25.

Your Home Buying Power

Give your home buyers money-saving answers to the most commonly asked questions on financing and tax advantages. Essential information on mortgage types, determining price range, deductions, and capital gains. $15 for set of 50.

New Home Buyer's Workbook

Give this workbook to your customers at the start of the job and see how easy custom work can be! Enclosed in a tasteful folder with room for your cover letter and business card, this workbook helps your customers make all their selections on time. Tips and suggestions are given at the beginning of each of the 7 sections. We have also included an extra set of forms, so customers can give one set to you and keep one for their records. $25 for set of 5.

Choosing Your Builder

Turn potential customers into buyers with this easy-to-read brochure. Concise and informative, it tells consumers how to make an educated choice when selecting a builder. $15 for a set of 50.

Make Your Move to a New Home Now

Tell your prospects why a new home is an investment for the future and the smartest purchase they'll ever make—and you will make sales! $15 for sets of 50.

To Build a Home

The Inside Story of Building the American Dream - An exciting 30-minute video.

An excellent video to have in your office, to show at schools or civic group meetings, to educate local officials, or to send home with a client.

With a home builder and a lively blueprint as hosts, and a supporting cast of hundreds of workers and tradespeople, *To Build a Home* shows just exactly what it takes to build today's new home.

To Build a Home follows the construction of an actual home from excavation to completion. Along the way, viewers gain an understanding of the home building process, from land acquisition and financing to site planning and home design, to construction to move-in. $20.

To order, call 1-800-223-2665

Code: QB